SUPPLY CHAIN MANAGEMENT FOR COLLECTION SERVICES OF ACADEMIC LIBRARIES

SUPPLY CHAIN MANAGEMENT FOR COLLECTION SERVICES OF ACADEMIC LIBRARIES

Solving Operational Challenges and Enhancing User Productivity

ZHENG (JOHN) WANG
Hesburgh Libraries, University of Notre Dame,
Notre Dame, IN

ELSEVIER

CP
CHANDOS
PUBLISHING
An imprint of Elsevier

Chandos Publishing is an imprint of Elsevier
50 Hampshire Street, 5th Floor, Cambridge, MA 02139, United States
The Boulevard, Langford Lane, Kidlington, OX5 1GB, United Kingdom

Library of Congress Cataloging-in-Publication Data
A catalog record for this book is available from the Library of Congress

British Library Cataloguing-in-Publication Data
A catalogue record for this book is available from the British Library

ISBN: 978-0-08-102031-9

For information on all Chandos Publishing publications visit our website at https://www.elsevier.com/books-and-journals

Publisher: Glyn Jones
Acquisition Editor: Glyn Jones
Editorial Project Manager: Thomas Van Der Ploeg
Production Project Manager: Debasish Ghosh
Designer: Victoria Pearson

Typeset by TNQ Books and Journals

CONTENTS

Part 3 Future Proof Library Supply Chain

PART 1

Introduction to Supply Chain Management and Case Studies

CHAPTER 1

Supply Chain for Libraries—Big Picture

PRELUDE TO THE NEED OF THE SUPPLY CHAIN

Many readers might wonder whether a whole book devoted to the topic of the Supply Chain is warranted for libraries. Is Supply Chain not a concept geared toward businesses, particularly in the sphere of manufacturing, retail, and logistics? Libraries are not-for-profit organizations, and therefore, they operate on a different set of values. So why do or should libraries care about Supply Chain or Supply Chain Management? Can a business process, such as Supply Chain, solve library challenges and serve library end users?

The author senses that the above questions might immediately emerge when library professionals see the title of the book. The answer is quite simple: the book's intention is not to set an agenda to turn libraries into for-profit organizations, but rather to introduce the concept to library colleagues as a means of serving their constituencies.

Because there are many different types of libraries, each offering many great services, the library type might determine the variety of services they may offer and how they might manage them. The focus of this book will be on the collection services of academic and research libraries only because collection services are bread and butter to academic and research libraries.

Supply Chain is not a panacea for solving all collection services problems, but it provides crucial insights into how collection processes flowing from vendors, to libraries, and then to end users can impact scholarship and research bottom lines. Supply Chain Management concept also offers a framework to guide libraries to get more mileage out of their collection budget and improve collection usage experiences.

So, if Supply Chain is the means, what is the problem that it is trying to help libraries solve? Let us look at the following series of examples set in an academic library setting from the perspective of subject liaisons, catalogers, library administrators, and, most importantly, patrons. The author believes that colleagues who have been working in the library setting must have found themselves in similar situations at some point in their careers.

Supply Chain Management for Collection Services of Academic Libraries
ISBN 978-0-08-102031-9
http://dx.doi.org/10.1016/B978-0-08-102031-9.00001-3

David was thrilled about his recent acquisition of a private collection of a renowned author that he was able to secure a couple of months ago. He knew his faculty and students have dreamed of getting hold of this resource for a long time. As the valuable collection arrived at his library's loading dock, he learned that he was out of luck in regard to available shelving space and available processing resources. Shelves have long become a highly competitive commodity among selectors after years of collection building. The university just started a conversation about a new storage facility to accommodate the ever-growing collection needs of the library, but it would not solve David's problem, which requires an immediate solution. The approval of creation of a new book storage building was not an easy decision to make at the university level as it came with a high price tag and competed with other institutional academic priorities. For now, David's library had to make a decision to backlog any incoming materials until the library comes up with a solution to accommodate more volumes. To make matters worse, even holding places for unprocessed materials at this point were becoming almost full. David was disappointed and started wondering when his faculty and students will be able to have access to this collection. This collection required much curation as it contained resources in a variety of formats, including cassettes, floppy disks, papers, pamphlets, monographs, journals, and two laptops from the 80s. He thought that perhaps catalogers and archivists could start processing the collection while the campus was solving the storage issues.

Jane, Head of Collection Processing, was panicked when David approached her with a request for processing the collection. She did not want to disappoint him, but she had few resources to deal with David's collection at that time. She was even not confident when her team would be able to get to his collection in the future because of other previously outlined priorities and commitments. Jane indicated that she was not made aware of his acquisition and thus had already promised other selectors to process their collections. She also hinted that she still had a lot of backlogs her team was currently plowing through, and thus, she had no idea when she could get to David's collection. Jane also indicated that her processing space was at capacity with recently acquired collections and that she was not even sure whether she would have space for David's collection. She suggested outsourcing the collection, but David was not so keen on the idea. David was concerned about costs and his time that would be required to oversee the inventory and quality assurance of the outsourcing job. As collection development was only part of his job responsibilities, he did not wish to have tasks related to outsourcing pull him away from his teaching and outreach assignments.

Director of the Libraries received a call from the Development Office concerning a collection grant received from a Foundation a year ago. The Development officer reminded her that it was time to report back to the Foundation on the grant's work progress. Lyn was listed as the legal person who signed the agreement with the Foundation and thus was accountable that reporting on the grant takes place. As readers may guess by now, the grant was to support David's recent exciting acquisition. Lyn learned from David that the library would not be able to meet the foundation's expectation to get the collection fully processed and made accessible by the date outlined in the grant. Outsourcing was not a viable financial solution because the grant was not big enough to sustain the cost. Under the pressures of current situation, Lyn decided to prioritize the processing of this collection in-house, which required the library to hold off on processing other collections. The processing priority became a politically charged topic. Lyn also decided it was time to start a librarywide collection deaccession project to create more room for incoming materials. Every liaison was mandated to become part of the project and was expected to manage expectations of the respective constituency. It is a tough path to communicate and align stakeholders' expectations.

The above stories are extreme scenarios that a library may occasionally encounter. Although those stories are fictitious, they might represent some real experiences of library employees with collection management. The introduction of the Supply Chain concept to library industry is an attempt to improve collection services by optimizing processes around them. As the above stories only cover some current challenges that libraries face, it is worth noting that Supply Chain can be very future-oriented, offering libraries efficient and economical solutions to align their collection services and processes based on user goals and social and technological contexts that exist in any given point in time. This chapter addresses the question of why libraries need Supply Chain by starting with examination of the current environment in libraries.

CURRENT CLIMATE

Those professionals who have been working in libraries for some time most likely experienced ebb and flow in funding from their "parent" organizations. In the United States, university libraries went through a very healthy growth time in the 50s to 60s of the 20th century. Libraries in Europe were also burgeoning around the same time period. When the new century began, support for libraries on academic campuses capped. In some rare cases, the funding increased, but for the most part, library budgets were either plateaued or cut. Budget cuts or budget freezes became a new normal in the past two decades. Most libraries have been dealing with the budget reality in a reactive manner that could be demonstrated as follows: spend less when budget cuts hit by acquiring fewer items, canceling packages, and reducing business and work hours; spend more when budget is restored by purchasing more, staying open longer, and hiring more.

However, such a passive reactive approach does not take into account the impact on user productivity, which remains constant regardless of the context of economic climate. A key question that library leaders should consider is how to sustain or even increase support of scholarship and research needs throughout good and bad times. Although many universities continue to grow in size and admission numbers and expand to add more disciplines taught and degrees offered, the percentage of library budget has not seen an increase—a dichotomy between the ever-growing needs of the academy and nongrowing or even shrinking resources in libraries. Therefore, libraries must manage their own operations with what they have using a proactive approach, while simultaneously advocating for more funding to match the growth of university and user needs.

STAKEHOLDER EXPECTATIONS

Another change since 2000 came in the form of growing expectations from library parenting organizations. Universities of today demand a set of new values, such as accountability and stewardship, to be explicitly expressed and followed upon, which may not have been prominent priorities previously. University asks academic departments and libraries to justify their funding levels by showing effectiveness and impacts, which in turn, become an evaluation or practice for securing future funding levels. Universities choose to do so as a way to prioritize funding opportunities because they also face constraints of supporting all the put forward aspirations and endeavors.

Influenced by the set of new expectations, libraries are trending toward assessing the relevance of their services and mapping them to learning outcomes. There have been many attempts to create a direct correlation between services that libraries offer and student performance. Although the relationship of libraries to learning results might be easily assumed, it is extremely hard to be quantified, unlike traditional metrics, such as number of volumes and size of collections. Nevertheless, libraries have been making efforts to demonstrate their contributions to campus learning and overall missions of their academies. For example, many libraries have been working on finding ways to engage and participate in research activities and academic teaching offerings.

As libraries come to realize that it is critical to adapt to the growing needs of universities and to embrace a new set of values, they must change their well-established and historically rooted operations under the current climate. Libraries have been gradually shifting existing resources from traditional services to emerging ones. Because additional resource support is limited, the velocity of the desired change is typically slower than it should be. So the best approach in current situations is to free up the existing resources quickly so they could be reallocated based on the institutional needs. Collection services are one of the most critical services that libraries offer to patrons. Not only collection budgets are tied to the service but also the majority of the current library human resources are allocated to the current collection service model. Even the library organization structure is often departmentalized based on collection processes.

The traditional model of collection services focuses on building collections. Collection building was one of the most meaningful tasks of the past to support the mission and goals of the academy. In fact, to this day, the collection size or the amount of investment dollars are two of the key metrics that benchmark prestige and ranking among peer institutions. Collection size is often representative of the value of organization's assets. At the same time, libraries often struggle to find funding needed to cover costs associated with their acquisition, curation, storage, and preservation. Based on the Goodhart's Law, when a measure becomes the target, it can no longer be used as a measure (Business Dictionary, n.d.). Libraries should really question the relationship of a large body of collections to user research and scholarly activities. In the current collection model, assessment of usage rate is not a primary indicator of success. It is a known fact that only a fraction of the entire collection body circulates. Taking the author's own institution's experience as an example, although the collection body continues to grow, circulation keeps dropping each year. Besides circulation, how do collections support end users? Or, rather, how effectively do collections support user needs? The answer to the above questions should be represented in a new set of questions that libraries ought to start exploring more. Goodhart pointed out the problem of assigning value to a specific variable to be used as an indicator (Business Dictionary, n.d.). Libraries are using the variables of size or investment as an indicator for the success of collection services. It is time to consider whether collection building is still one of the most meaningful means to support the academy.

The cost of sustaining the current collection services model is often significant. Many libraries have gone through phases finding ways to expand limited collection spaces. Those institutions who have gone through the exercise understand the difficulties to secure funding for additional collection storage spaces, a need competing with other university priorities. Physical space will be an even more scarce resource in the future, and the collection storage conversations with university administration will become even harder over time. Having to deal with the aftereffects of congregating large bodies of collections, many libraries took steps to collaboratively develop shared collection facilities, both analog and digital storages, as a solution aimed at addressing packed collection spaces. Although shared facilities is the direction libraries should move into, institutions also need to face the fact that there exists a high level of collection duplication across the board because of the long-standing practice of collection building at each institution. With that in mind, libraries should critically look at the value of overlapping collections and their opportunity costs to their organizations. As long as the size of the collection or expenditures associated with it remains the main focus, not many libraries will be serious about considering changes and evaluating the deep implications of the current model on their needs to align goals in support of their academies' changing needs and priorities. Although some institutions have already started to shift their collections focus on rare and special collections and cater toward more research and learning needs, the change is not systematic enough at each discipline, institution, and association level.

SERVICE ORGANIZATION WITH LEAN RESOURCES

Libraries, first and foremost, are service organizations that fulfill several functions in society. Universities rely on libraries to advance teaching, learning, and research missions. Libraries of today are also organizations with lean resources and high expectations of meeting growing needs of their parent institutions. Library nature of service orientation requires them to share service design and development with all academic departments. Because resources are few and needs are many, libraries should seek the most economical and efficient ways to maximize the utility of their resources to support university missions. Because libraries exist in a less competitive and more relaxed and laid-back environment compared with for-profit organizations, libraries often do not operate in line with the full "Lean" concept, which will be covered in Part three—Future Proof Library Supply Chain of the book. Operational precision is often not as a big of a concern as services themselves. Libraries seldom promote or advocate the relationship between process excellence and service excellence. So, in reality, process management is loosely coupled with services where either processes are neglected, or they are designed without the context of service results. Because procedures have long been established within departments, innovation of processes across departmental lines is challenging because of deeply rooted traditions and because processes that have been in place work well under immediate conditions. As libraries reflect on their social responsibilities to user communities, process improvements would provide additional ways to realize their organizations' missions. The three-level library maturity model discussed in the following sections demonstrates the correlation between process excellence and service excellence. The maturity concept is formulated based on the assumption that libraries are lean service organizations.

Level One: Departmental Productivity in Processes, Metrics, and Services

Individual departments are owners of their processes. Services are managed at the departmental level. Each department defines, improves, and measures its internal processes. Each department is responsible for any functional outcomes.

For collection services, subject liaisons are responsible for collection development, expenditures and size of the collection as measures of their performance. Acquisitions are responsible for ordering, shipping, receiving and checking in collections; and their performance is often measured by the number of orders placed and items received. Cataloging is responsible for describing collections, and thus, their performance is measured by the level of adherence to professional standards of records. Circulation and stacking are responsible for collection services, while level of responsiveness and service satisfaction measure their performance.

At level one, process coordination among departments is ad hoc and inconsistent. Libraries have no standardized framework to develop and improve cross-departmental processes that contribute to well-defined service outcomes. Service success happens in isolation at the departmental level, which is often difficult to translate to user success.

Level Two: Organizational Productivity or Service Productivity in Process, Metrics, and Services

Libraries are owners of their processes. Services are coordinated at the organizational level. All departments participate in defining, improving, and measuring cross-departmental processes. Libraries are accountable for service outcomes.

At level two, process improvements are service oriented; however, service performance is not mapped out to patron success. Service outcomes remain library-centric or service-centric. For collection services, collecting is the main service goal, and thus, libraries design processes around building up the body of their collections. User services are by-products of collection building activities.

Libraries need to understand the fact that the above-described library-centric approach worked well and was more aligned with scholars and researchers' goals when publishing was analog only. Before the Digital Age, production of publications was rather limited compared with what we witness today. Geographically imposed limitations of timely delivery or sharing of print collections among libraries also required libraries to invest effort and resources into building their own collections to support campus scholars and researchers.

Level Three: Patron Productivity in Process, Metrics, and Services

Patrons and libraries are co-owners of processes. Services are coordinated at the organizational level. Patrons participate in defining, improving, and measuring processes. Libraries are accountable for optimizing processes that support patron productivity.

At level three, libraries deal with misalignment between the organizational and service goals and strategies and those of their users. Libraries set up consistent patron-oriented processes across the whole organization. Libraries also establish systematic metrics for assessing and improving their process performance and their contributions to user outcomes and expectations. Libraries focus on continuous improvement, change management, and innovation to maximize their value to user productivity.

A service organization keeps its patrons at the core of what it does even as its context evolves. A service organization always transforms its processes, services, and sometimes itself to support user aspirations by using available resources.

From the above perspective, many libraries of today are perhaps in transition from level two to level three. It is possible that libraries have already aligned several processes with those of their users and established a few services that address specific user problems. Patrons have become the main beneficiary of service improvements; meanwhile, libraries continue to hold on to some processes and activities that are still very much library-centric and were originally designed for analog collections only. Libraries, as service organizations with lean resources, must strategize about how to use their limited resources to support ever-expanding service demands.

DIMINISHING PURCHASE POWERS

As indicated at the beginning of the chapter, libraries went through a growth period. Although libraries should continue growing in the areas of developing research and scholarship, their budgets are not rising in proportion with the demands. In some ways, it is even worse because library budgets are shrinking in comparison with what they could afford to buy then and what they can buy today. Although some libraries experience budget cuts from time to time, most libraries' budgets are shrinking primarily because of their diminishing purchase power. Retrospectively looking at the past 50–60 years, universities increased library funding overall; however, libraries are constantly finding themselves in a position of looking for funds to cover inflation adjustments on collection purchases or licenses. It is quite a common practice that any increase of library budget is often used to reconcile the disparity that results from continuously rising fees for existing collections. This means that libraries do not have dollars to manage strategic growth areas and instead have no choice but to resign to simply keeping the current level of services.

As the inflation rate keeps going up each year, it is rather impossible for libraries to keep their current operating model if they aspire to develop additional areas of expertise. As collections are getting more and more expensive over time, libraries will eventually find themselves in dire situations trying to balance current services and future needs. Another meaning of the word "lean" is fit and healthy. Libraries can invest in many of their operational areas to make them much "healthier." Under this context, Supply Chain is a perfect discipline that may help libraries carefully examine their workflows, processes and procedures to drive higher efficiency and productivity. The Supply Chain has potential to maximize the utility of library resources to best meet patron needs.

USER CONTEXT SWITCH

Before embarking on a journey to learn Supply Chain concepts and its implementation, the chapter needs to discuss key characteristics of majority of library patrons. Libraries, as service organizations, need to respond to their patrons' needs quickly. Library patrons have "migrated" into the digital world, which came with new information-seeking habits. Those habits are different from those of the analog world, although analog still dominates certain demographic populations and academic disciplines.

The Internet has completely transformed patrons' activities. Online is not a specific lifestyle or choice, but an absolute core part of people's everyday life. The Web became a primary platform for learning and teaching. Library patrons now enjoy more convenience and are given more options to "shop" for knowledge and information online. More and more patrons choose to interact with libraries in the virtual world. Internet granted instant gratification to users, as if the content that is not discoverable online does not exist at all. Library's ways of organizing and delivering content have started showing their age. It is about time to examine the collection management processes. Instead of locking their collections, libraries should focus on delivering more content to their patrons directly over the Web, as the Web has been a default platform for disseminating knowledge.

REFERENCE

Business Dictionary. (2017). (n.d.). *What is Goodhart's law? Definition and meaning*. Retrieved from: http://www.businessdictionary.com/definition/Goodharts-law.html.

CHAPTER 2

Emerging Values

BUSINESS VALUES IN LIBRARIES

Chapter 1 provided an introduction and overview of what Supply Chain can offer to libraries in regard to utilization of limited resources with the goal of creating maximization of user productivity. It is important to remind readers that Supply Chain is a business process. There still might be ideological resistance to the adoption of a business concept in not-for-profit organizations. It is necessary to have this chapter devoted to the convergence of business and nonprofit organizations, particularly emerging value that was transferred from business and adopted by nonprofit organizations. It is true that business world can learn much from nonprofit organizations, but in this book the author will focus on the reverse idea, namely, what nonprofits can learn from business organizations.

It is true that libraries do not perform based on the same principles as for-profit companies, but libraries have been borrowing business ideas and concepts from the business industry for quite some time. Such terms, as subtle as customers, stakeholders, and product, and as controversial as Return on Investment, have frequently appeared in library literature. Hayes and Brown (1994) indicated that the concept of Total Quality Management (TQM) stormed into libraries as far back as some 20 years ago. As TQM made its way from Japan to the United States, libraries found insights into financial relationships and organizations from the writings of W. Edwards Deming (Neave, 1990; Walton, 1986). The author understands that Deming's philosophy is about creation of relationships among units of an organization, so that they may work together in the most efficient manner to deliver what customers need. In other words, it is a customer-driven methodology that drives collaboration of otherwise siloed units to provide better products and services to end users. In TQM, all members of an organization participate in improving processes, products, services, and the culture in which they work (American Society for Quality, n.d.). So, it is implied that Supply Chain is worth some attention garnered from libraries and requires all members of an organization to participate in its implementation. Part three — Future Proof Library

Supply Chain Management for Collection Services of Academic Libraries
ISBN 978-0-08-102031-9
http://dx.doi.org/10.1016/B978-0-08-102031-9.00002-5

Supply Chain of the book will introduce several more Deming's ideas of framework for the adoption of the Supply Chain. Hayes and Brown (1994) boldly claimed that "if one were to map the financial relationships and examine the energy that goes into those relationships, it would be clear that the libraries is a business and will continue as a business for the foreseeable future." Since then, libraries have been borrowing business models for implementation in their industry. As indicated in Chapter 1, universities increasingly expect units on campus to demonstrate their impact through a set of business values.

So, then, are libraries businesses? Many professionals would probably find the term to be borderline offensive and in contradiction to the core belief why they work in libraries, in the first place. One may argue that there are substantial differences between the mission of a business and that of a not-for-profit organization. There is not much literature that supports the claim. The author agrees that the term business has the association with financial bottom line rather than that of a noble cause of nonprofit service organizations. One must understand that many business concepts were created under the circumstances different than those of the libraries. Simply shoehorning them into library environment would be painful and highly counterproductive. Business ideas will not work for libraries until first they honestly examine their own operations and unique characteristics.

The author believes that there is a false dichotomy between the terms business entities and not-for-profit organizations. In light of the philosophy that exists for both types of entities, each has to exist to serve their users. To accomplish this mission, each has to offer products or services that must be valuable to users. The products or services must solve user problems or add benefits to the already existing ones. The differences between the two are the financial models and environments they operate under. Businesses have to make money to sustain, compete, and thrive, whereas libraries receive funds to run their operations. Business is highly competitive, whereas libraries are very collaborative by nature. Regardless of the core differences between for-profits and nonprofits, libraries and business must use their resources well.

BUSINESS VALUE IN COLLECTION SERVICES

All resources initially come in the form of budgets. Every library's budget contains a large collections budget. Even the labor pool of the budget portion connects to collections for the most part. This section of the chapter will discuss resource use and allocation to demonstrate the importance for the service to embrace Supply Chain. From the aspect of financial management, libraries are just like businesses. It is critical to understand collections from the budgetary perspective.

It is common that collection funds are often compartmentalized in silos. Except for finance department employees and some senior managers, not many understand how individual pieces of collection funds fit into and connect to the whole picture. Selectors have their allocated budget funds to build collections. Technical Services have a separate budget for managing acquisitions, cataloging, shipping, and preparation. Conservation and Preservation have an exclusive budget for collection repairs and long-term care. Circulation has their budget for document delivery, interlibrary loans, and other incidental costs associated with collections, such as maintenance of bookshelves.

Library administrators need to look at the overall collection budget model and selection processes. Because of the lack of a systematic approach in collection funds, departments usually steward accounts exclusively at their own levels, not taking into consideration how it connects to budgets of others in the organization or those of the entire system. Selectors try their best to expend their allocations each year. Catalogers are not concerned with the cost of collections as long as each item has a record to show its existence in the catalog. Conservation and Preservation budget is seldom large enough to match the speed of acquisition and increase in the body of content. Regardless of the size of the collection that libraries purchase or license, Interlibrary Loan still has to borrow titles to satisfy patrons' demands. The current budget practice created a phenomenon of what could be called "the left hand does not know what the right hand is doing." It also caused many process issues that exist in a somewhat disjointed environment. As the budget for human resources is not coordinated throughout the whole organization, it is very common that libraries often hold backlogs of collections to be processed. Of course, collection services also include many other aspects besides finances. Financial perspective alone cannot determine service quality. The intention here is to demonstrate that even financial practices have impacts on the overall service model and processes. If the profession could embrace finance management as a business management piece, why not open up for other meaningful business processes that may also improve service results? Financial issues are seldom the only challenges that libraries face.

As libraries are buying greater variety of collections consisting of books, serials, article databases, and data sets, they have to plan for allocation of their funds to purchase and curate expanding content and their formats. Coping with price increases, inflationary rates, and format evolution is not a new problem that libraries are set to tackle. The issues contributing to difficulty in balancing a limited budget to manage the lifecycle of collections from access to long-term curation have been with the libraries since publishing industry moved beyond producing content exclusively in paper format. The argument then is if libraries have been spending that much energy, effort, and money to catch up with the format updates and cost increases for a very extended period and still have not caught up, why would not libraries consider alternative options to fundamentally change their position from reacting to anticipating changes.

Since the latest economic downturn in 2008, higher education has adopted a new approach, "do more with less," as a reactive measure to deal with the current climate of financial difficulties. This phrase is another business slogan that perhaps has widely spread in nonprofit sectors. People may interpret the term very differently, which could run the full gamut from putting in long hours to implementing layoffs, which goes against the traditional values of work–life balance and employment stability present in service organizations. The economic downturn of 2008 has led to reduced staffing levels in libraries (Nicholas, Rowlands, Jubb, & Jamali, 2010). The author understands peers' suspicions and concerns of adopting any business-born concepts into the library world. Certainly, for collections services, "do more with less" remains merely a slogan rather than a reality because of the resources required to sustain the current collection model. The purchase cost of academic journals has risen significantly more than the general inflation increase in the economy and the budgets of libraries. It has been causing many libraries to spend significantly more to get less. The large increase in the cost of journals in relation to the cost of books during the period 1886–2001 reached 227% for journals and 75% for books, compared with general inflation in the United States, according to data statistics from the American Research Libraries (Katsirikou, 2011). Regardless of how "do more with less" was interpreted and implemented, the phrase means improvement to their management side of operations and running services more efficiently than before.

Since 2008, the author also observed firsthand more adoption of the business concept of "just-in-time" and its application to the collection development over the traditional "just-in-case" model in libraries. "Just-in-time" is one of the elements of the Supply Chain concept, which Parts two — Library

Supply Chain Model and three — Future Proof Library Supply Chain of the book will cover in detail. Traditionally, libraries' missions have been rooted in collection building, and thus, library budgets have been prioritizing materials purchasing and growing physical collections. Going as far back as the 1990s, "just in time" or "just in case" models were offered as a solution to address competition between monographs and serial subscriptions for the same pool of collection budget (Hayes and Brown). Academic libraries' acquisitions began to face financial difficulties in the mid-1980s, which influenced collection development. Studies have shown that serial unit costs have grown by 180% and serial expenditures increased by 132% in the past 20 years (ARL statistics), which lead libraries to look for alternative solutions to the previously established subscription funding model (Katsirikou, 2011). As libraries experience more pressure induced by budget reality of today, some of them started to lean toward more customer-oriented acquisitions rather than continuing their "just-in-case" collection building activities. Currently, most libraries set up a separate budget to manage just-in-time collections. The book proposes that libraries take that approach even further. The "just-in-time" concept will be developed by the author into a collection Supply Chain strategy called "Make-to-User" in Part two — Library Supply Chain Model of the book.

The Supply Chain concept also offers library an opportunity to examine the "total cost of ownership" of their collections. "Total Cost" perhaps is another business idea that libraries could borrow from the for-profit industries. The term gives library professionals a new perspective to understand how much collections cost beyond the initial acquisition. As collection budgets are often limited to accounting for the cost of publications (referring to the way how budgets are compartmentalized), it is crucial to consider total collection costs that also include shipping, labor, supply, equipment, furniture, facility, and utility costs to manage and curate the acquired collections. For some institutions, it was easy to make decisions on their collections based on the Total cost of ownership; however, for many libraries who center their priorities around collection building, the concept has little effect on guiding collection development activities. The crux of the matter is deciding what the goal of collection services is. Is it the size, the completeness, the depth, the preservation of the collections themselves, or the collection's usefulness and contribution to user productivity? Do many libraries find it difficult to have conversations around the overall cost of collection services? What is the total cost of keeping an item in perpetuity, if it never circulates or even if it circulates once or a few times only? What are some viable alternatives that might cost libraries less to deliver the same value to end users?

Return on investment (ROI) is another term that the library industry has been embracing. The concept closely links to the value of accountability and stewardship of resources. Library administrators and university leaders are looking for effective ways to use funds and efficient processes that enable daily operations. The term could be viewed as controversial in the environment when libraries explore efforts to quantify their impact on institutional priorities and missions in the face of budget challenges and the upper management questioning of libraries continuing relevance to the academy in an all-digital information world (Neal, 2011). The author agrees with Neal that the application of the term in quantified financial measurements of library successes is problematic and cannot agree more on the point of understanding of ROI from the user perspective. Neal (2011) advocated for starting with understanding who users are and what they need. Customer-centric processes are core to the Supply Chain model. Parts two — Library Supply Chain Model and three — Future Proof Library Supply Chain of this book will focus on processes that originate from end users, traveling back to library processes and then vendor ones, under the idea of the Extended Supply Chain. Perhaps, ROI application should be expressed in the Return in Value to users rather than in quantified financial returns.

Library stakeholders desire to see the funds allocated to libraries return value and have a direct correlation to the mission and priorities of the institution and the academic success of students, faculty, and researchers. It is not an expectation that targets only libraries, per se, but one for all departments on campus. As many business concepts and methodologies have been penetrating the higher education setting, library professionals should take this as a signal of shifting culture. Although many are still searching for how exactly some of those business concepts apply to higher education, and some concepts might never find successful application in nonprofit organizations, libraries should not discredit the core value of the terms described in the chapter and must continue to seek ways to improve libraries value to end users' aspirations. Neal (2011) pointed out the importance of sensitivity to new measures that are very different from those that have governed library thinking for many decades. Libraries must match their operations with the current environment. Libraries are still the same service organizations and are in the same business to help people, regardless of changing environment. Libraries must search for new meanings of what it takes to continue in the business of helping people and what it means to be productive, efficient, and effective service organizations.

Neal explains that library success does not equal resource allocation, but instead resource attraction. He sees value defined as quality plus traffic; quality equals content plus functionality. Applying his definition to collection services, libraries should consider how the allocation of collection budgets creates attractive collection services. How then libraries may improve the quality of collection services? It is clear that besides the collections themselves, their functionality and traffic (collection utility) also contribute to the overall quality of collection services. The Supply Chain features actually can transform collection content. Supply Chain can help libraries transform other services as well, such as research data support and consultation services. Part two — Library Supply Chain Model of the book will concentrate on introduction and application of the Supply Chain principles to library industry and its collection services. Part three — Future Proof Library Supply Chain of the book will focus on ideas for adoption and implementation considerations. The Supply Chain Models for support of other services will be different than those for collection services. The book's goal is to dive deeper into the area of collection services, yet there are many additional topics that could be covered as part of Supply Chain methods; hence, the author encourages peers to explore other possible models for all library services.

REFERENCES

American Society for Quality. *What is Total Quality Management (TQM)?* http://asq.org/learn-about-quality/total-quality-management/overview/overview.html.

Hayes, S., & Brown, D. (1994). The library as a business: Mapping the pervasiveness of financial relationships in today's library. *Library Trends*, *42*(3), 404–419.

Katsirikou, A. (2011). *Open access to STM information trends, models and strategies for libraries (IFLA publications 153)*. Berlin: De Gruyter Saur.

Neal, J. G. (2011). Stop the madness: The insanity of ROI and the need for new qualitative measures of academic library success. In *Association of College and Research Libraries National Conference* (pp. 424–429). Philadelphia: Association of College and Research Libraries.

Neave, H. R. (1990). *The Deming dimension*. Knoxville, TN: SPC Press.

Nicholas, D., Rowlands, I., Jubb, M., & Jamali, H. R. (2010). The impact of the economic downturn on libraries: With special reference to university libraries. *The Journal of Academic Librarianship*, *36*(5), 376–382.

Walton, M. (1986). *The Deming management method*. New York: Perigee Books.

CHAPTER 3

Supply Chain and Supply Chain Cases

DEFINITION OF SUPPLY CHAIN

Supply Chain is a relatively new concept to libraries, and hence, this chapter provides a general introduction to the idea, its definition, history, goals, and a couple of case studies that are relevant to libraries.

What is Supply Chain? Some people might consider it to be the same as logistics. In their 1999 article, Lummus and Vokurka described what supply chain was not. They cited a number of terms that people use interchangeably with Supply Chain, including logistics. Although logistics encompasses flow of products or services from suppliers to customers, Supply Chain is a significantly broader concept.

Many scholars (Ayers, 2000; Blanchard, 2010; Lummus & Alber, 1997) in the field have published variations of the Supply Chain definition. Although it is pointless to debate which definition is better or more accurate, it is critical to understand how each fits the goals of particular groups.

The Oxford English Dictionary defines the Supply Chain as "(a) (collectively) the routes or means by which supplies (esp. those for a military force) are received; (b) Business the chain of processes involved in the production and distribution of a commodity" (Oxford University Press).

Lummus and Alber (1997) interpreted it as, "the network of entities through which material flows. Those entities may include suppliers, carriers, manufacturing sites, distribution centers, retailers, and customers."

Quinn (1997) presented it as "all of those activities associated with moving goods from the raw materials stage through to the end user, including sourcing and procurement, production scheduling, order processing, inventory management, transportation, warehousing, and customer service. Importantly, it also embodies the information systems so necessary to monitor all of those activities."

Wikipedia (2016) explained it as "… a system of organizations, people, activities, information, and resources involved in moving a product or service from supplier to customer."

Supply Chain Management for Collection Services of Academic Libraries
ISBN 978-0-08-102031-9
http://dx.doi.org/10.1016/B978-0-08-102031-9.00003-7

Based on the preceding definitions, one can reason that the Supply Chain is much more complex than simply moving goods or services to users. More importantly, it systematically directs the flow of information, knowledge, and human and financial resources in concert with the flow of products or services to users.

Notwithstanding the variations of the definition, one consensus on the Supply Chain among the professionals was "… all that happens to a product or service from dirt to dust." Ayers (2000) particularly emphasized that the supply chain's purpose was "to satisfy end-user requirements with products and services from multiple linked suppliers."

In this sense the Supply Chain extends from vendors' providers to customers' customers. In the publishing, knowledge, and information industry, the Supply Chain may begin with tree farms, papermaking factories, silicon mines, and networking and computing technology companies to publishers, content aggregators, book retailers, library patrons, used-book sellers, and recycle services.

Libraries are part of a super "Extended Supply Chain." When viewed through this lens, library professionals might be able to see their challenges from a slightly different perspective that libraries, publishers, and vendors are all responsible for delivering value to customers. The win–win mindset is in the best interest of any party on the Supply Chain. Openness, transparency, and collaboration must replace confrontational maneuverings and antagonistic tactics and strategies to create a cohesive and effective system that can solve shared supply chain problems. Additionally, the concept of the "Extended Supply Chain" reveals that libraries, publishers, and vendors are components of a larger ecosystem. They are the "links" that connect products to customers. Just as other businesses that are situated in the middle of their Extended Supply Chains have transformed to redefine, readjust, and adapt to technological and social changes, the publishing, knowledge, and information industry is also progressing through a similar phase of maintaining relevancy to patrons. It is crucial for the industry to realign with the emerging civilization and needs of patrons; otherwise, it will inevitably disappear just like the "links" in other industries that did not change.

Let us take a much closer look at the Supply Chain in the context of libraries. The complexity of library processes qualifies for applying supply chain theory into everyday practices. Libraries are not a single entity that exists alone and only creates knowledge and packages it to users on their own. No one in the current climate would think of libraries as a small family-owned store in his or her neighborhood. Libraries are a complex system with defined processes and procedures. Libraries depend heavily on many other vendors and peers to support their users.

The fundamental goal of the supply chain is to meet end-user requirements and satisfy user needs. Analogous to the mission of the for-profit sector, the mission of a library is to serve the needs of scholars, researchers, and learners. The end-user perspective in the Supply Chain is a critical notion for libraries. Although companies are trying to differentiate based on their physical products, more and more functions of the products are very much alike. Similar to products of other industries, library products are functionally similar. For example, it does not matter what formats of collections libraries acquire, the function of their collections is in their utility to patrons. The fact is that libraries have no control over the formats in their collection. Therefore, how might libraries set themselves apart from other information providers? Just as many businesses now design products originating from their customers' ideas, a backward scan along the supply chain from the patron's library can also identify numerous opportunities from access services, technical services, and acquisitions methods to create better alignment between user requirements and the library's products. Companies often differentiate substantially by applying the notion of "extended products." For example, what is the type of user experience that credit card companies offer to their cardholders when they call customer support on the phone? User satisfaction relies heavily on how those credit card companies handle those calls. The call center experience becomes a fundamental element of the "extended product" that differentiates credit card companies. Similarly, libraries can offer unique "extended products" based on their collections. The concept of Supply Chain offers libraries a framework to engineer appropriate and productive research experiences for their patrons.

The Supply Chain requires libraries to be concerned with not only the quality of their collections but also their acquisition methods, preparation procedures, discovery routes, delivery options, and all other possible factors affecting collections as a service, such as preservation, facilities, and IT systems. Unfortunately, aside from the quality of the collections, other aspects of collections often get only cursory attention in libraries. Normally, any product or service has a life cycle, which includes both the market cycle and the usage cycle. For libraries, the market cycle is similar to many businesses, but their products are nearly "imperishable" compared with those of other types of operations. Libraries must support the usage cycle of their products over an extended period. Other businesses might have the option to "retire" previous models without offering further support or to simply toss goods that expired. Libraries must develop excellence in critical expertise, such as preservation, collection facility management, data storage, and IT, to adequately support the long life cycle of their products and services.

Another intriguing aspect about the Supply Chain is its bidirectional flow, which is unlike many other concepts, such as logistics. The two-way flow means sharing information and local processes both among internal departments and with external vendors. Any improvement attempted to address local processes will have limited impact on users. Within the current state of libraries, many local operations are self-contained. Flows of information, financial decisions, and resources seldom go downstream and rarely run upstream. The problem with this model is that many decisions are made based on the priority of a group rather than of the entire organization. For example, it is likely that cataloging priority is not the same as acquisition priority, or collection development procedures do not have the same priority as that of the preservation and conservation group. This silo "functional" practice is profound along the "Extended Supply Chain" as well. For example, annual subscription adjustments make complete sense for publishers or content aggregators; however, as we learned in Chapter 1, many library budgets are flat if not slashed. These types of adjustments are rather unsustainable and eventually will hurt publishers and content aggregators themselves. When libraries were squeezed to adjust to those kinds of hikes, it was evident that they either cut subscriptions or switched to alternatives.

Libraries must improve the information and financial flow of their supply chain. Often, libraries prioritize the physical flow of their products. Again, look at the collection as one example. Libraries have designed their organizational structure to move collections from one department to another, from one team to the next, and from one individual to another. What is missing from the process is the knowledge of the criticality or priority of the collection to the institution or its patrons and the costs associated with each step of the process, such as creation of catalog records. This is not to say that such information does not exist, rather it is contained in a department or with key personnel. Without the understanding of such information, each department, team, or individual can only prioritize its own work. Such an organization leaves room for inefficiencies and errors along the Supply Chain which, thus, ultimately impact collection services to patrons. As volumes are traveling through departments, there is value for people involved to have the entire picture of financial and other information about the collections. Doing so will cultivate ownership of the process, and, hence, create opportunities for innovation and improvements for the entire chain. The literature shows that other industries have also experienced similar silos in the past. Ayers (2000) described the phenomenon, "companies are a collection of individual departments... Each department has, to a large degree, its agenda. Oversight of links between departments is weak within the company. Between companies in the supply chain, it is practically nonexistent. Most improvement initiatives are local. They may or may not improve the supply chain as a whole. In fact, a "local optimum" may be injurious to the whole, yet department heads take credit for their "savings"." The use cases section of the chapter will demonstrate that leaders in business have departed from the silo structure to a more integrated and collaborative style to drive value for users.

With the "supply chain" defined, it is time to look at the "Supply Chain Management (SCM)." Council of Supply Chain Management Professionals (CSCMP) explained SCM as follows:

> *Supply Chain Management encompasses the planning and management of all activities involved in sourcing and procurement, conversion, and all logistics management activities. Importantly, it also includes coordination and collaboration with channel partners, which can be suppliers, intermediaries, third-party service providers, and customers. In essence, supply chain management integrates supply and demand management within and across companies. Supply Chain Management is an integrating function with primary responsibility for linking major business functions and business processes within and across companies into a cohesive and high-performing business model. It includes all of the logistics management activities noted above, as well as manufacturing operations, and it drives coordination of processes and activities with and across marketing, sales, product design, finance and information technology.*
>
> ***Wikipedia (2016)***

The APICS Supply Chain Council summarizes the concept of the SCM as "plan, source, make, deliver and return" (Blanchard, 2010). CSCMP summarized "Supply Chain Management integrates supply and demand management within and across companies" (Wikipedia, 2016). Ayers (2000) pointed out that the goal of SCM is the "design, maintenance, and operation of supply chain processes for the satisfaction of the end user needs."

Although it is not an established research area, nor an organized practice, in Information Science or Library Studies, because librarianship is a discipline centered on knowledge acquisition and dissemination, SCM is a natural fit to optimize library processes across departments and vendors to satisfy patrons' teaching and learning needs. This book will establish SCM practices for libraries and suggest how libraries may operationalize, maintain, and improve SCM along the way. Along with many established processes in the libraries, topics such as process transformation, efficiency, and effectiveness will be explored as well. The next section of the chapter will demonstrate how scholars, researchers, and practitioners in many fields have recognized SCM over time as a discipline, similar to marketing, communication, and finance.

HISTORY OF SUPPLY CHAIN MANAGEMENT

SCM started at the dawn of commerce (Blanchard, 2010) with relatively straightforward and labor-intensive processes (Robinson, 2015). There was not a particular moment that proclaimed the birth of SCM, but it has been an evolution over the past 100 plus years. The history of the supply chain initiative can be traced to early beginnings in the textile industry with the quick response program and later to efficient consumer response in the grocery industry (Lummus and Vokurka, 1999). SCM becomes much more popular in 1980 and through 1990 when firms saw the benefits of collaborative relationships within and beyond their own organization (Lummus and Vokurka, 1999; Robinson, 2015). Study of the history of SCM will illustrate its connections, applications, and relevance to the library field and provide a frame of reference to the current environment as libraries look for ways to implement the idea.

A few milestones are worth mentioning in SCM history. The Industrial Revolution gave rise to industrial engineering and operations research. The most prominent person was Fredrick Taylor with his "The Principles of Scientific Management" back in 1911. Since then, people have examined logistics as a means to improve productivity; however, its impact on a company's bottom line was less evident to owners. Logistics were ingrained with everyday activities as the processes became much more complicated with the improvement of products.

By the 1940s, during World War II, the value of operational efficiency rose to prominence as it determined lives of soldiers and outcomes of battles. Extensive research focused on examining complex war logistics problems was conducted. A new term, "Supply Chain Engineering," was emerging in many trades. This time was a boom for research in the area; however, customers were not part of this picture. The improvements were addressing internal operational inefficiencies. Studies centered on warehousing and workforce processes, but they began to include means of transportation and distribution as products were being hauled from factories to store fronts.

Academic research and education about Supply Chain trended around the 1960s to meet the increasing needs of many enterprises. The rise of computerization at the time transformed the operation of the Supply Chain, particularly in warehousing and inventory controlling. Computing technologies brought supply chain theories into practice through the late 1970s and early 1980s. For example, academic units such as Supply Chain & Logistics Institute of Georgia Tech (Supply Chain & Logistics Institute, 2015) were created to further scholarly research. This resulted in more technological innovations into the discipline and field operations.

The 1980s and 1990s marked increasing industry awareness of SCM as a possibility to significantly improve the bottom line and competitiveness, and, therefore, businesses invested in professional training and new technology. Currently, SCM is a force for transformation and a discipline of competitive advantage. Ellram and Cooper (1993) considered SCM as "an integrating philosophy to manage the total flow of a distribution channel from supplier to ultimate customer." Monczka and Morgan (1997) believed that supply chains, not firms, compete and that those who want to be the strongest competitors are those who "can provide management and leadership to the fully integrated supply chain including external customer as well as prime suppliers, their suppliers, and their suppliers' suppliers."

The essence of the discipline has gone a long way from workflow efficiency, material handling, warehousing, and inventory control to a systematic and strategic coordination of business functions that add value to end users within and beyond the boundaries of an organization. The practice has evolved from pressing down costs to driving up value, from internal goals to customer aspirations, from delayed fulfillment to near–real-time delivery, from silo to collaboration, and most importantly from bounded tactics to cross-industry leadership. The name change of the organization of the Supply Chain professionals is evidence of the previously described transformative shift. The National Council of Physical Distribution Management formed in 1963, later became the Council of Logistics Management in 1985, and was last renamed to the CSCMPs in 2005 (Robinson, 2015).

Generally speaking, libraries are still in the early days of adopting SMC. Although there have been considerable technological advancements in the library field, the holistic improvement of coordinated cross-departmental and cross-vendor processes has been mostly missing. Just as many companies have been leveraging SCM to reconstruct themselves, libraries need to capture this moment to reevaluate and reexamine their current operations in the context of the Internet, Social Media, and other technological advancements. SCM can add meaningful benefits and advantages if libraries are willing to adopt and build on its concepts.

GOALS OF SUPPLY CHAIN MANAGEMENT

Joel Sutherland of J.B. Logistics pointed out succinctly, "Supply Chain Management is the systemic, strategic coordination of the traditional business functions and the tactics across these business functions within a particular company and across businesses within the supply chain for the purposes of improving the long-term performance of the individual companies and the supply chain as a whole" (Ayers, 2000). SCM should achieve efficient fulfillment of demand, drive outstanding customer value, enhance organizational responsiveness, build network resiliency, and facilitate financial success (CSCPM, Chen, Defee, Gibson, & Hanna, 2013). This section will discuss the goals of SCM in the context of libraries.

Fulfillment addresses inventory management, and, ideally, inventory levels according to demand. In the for-profit world, types of products and the amount in stock are the keys to fulfillment. It is best practice to keep the merchandise readily available, neither out-of-stock nor overstocked. This level of efficiency will maximize the bottom line and free up cash assets for other critical needs. Fulfillment might be used as a term in Interlibrary Loan service, but it does not have the same meaning as that of SCM. Libraries could apply the SCM fulfillment concepts and practices in their collection services, although it can be a very controversial topic. Simply stated, SCM fulfillment is not what academic and research libraries have been doing as an established process; what they have been doing is to "overstock" collections, just in case. In a later chapter touching on innovations, we will discuss more about SCM fulfillment and academic libraries.

The ultimate goal for SCM is to drive outstanding customer value. This must start with a clear understanding of what the customer needs. Supply chain strategies, design, and capabilities should emanate from customer requirements (Sweeney, 2011). The result will be higher quality services, reduced service variability, and fewer service exceptions. The value to customers should also drive how fulfillments should be set up. Libraries have adopted several customer-centered methodologies over time. For instance, with the popularity of library websites, personas and user-centered design methods have made inroads into certain functions of libraries. Taking advantage of the same trend, libraries could model their SCM based on user behaviors mentioned in earlier chapters to drive cross-institutional values for their patrons.

Responsiveness and resiliency arise from the fact that SCM creates lateral changes by cutting through existing departments for collaboration and coordination. SCM can help organizations adapt to the challenges of globalization, economic upheaval, expanding consumer expectations, and related issues (Coyle, Langley, Novack, & Gibson, 2013). Responsiveness and resiliency are even more relevant to service organizations, as timeliness reflects the quality of services and resiliency enhances the sustainability of services and the long-term return on value for users. Economic upheaval has been impacting library budgets during the past decade. Patrons in the Digital Age have different expectations than those in the print world. Although globalization presents challenges to libraries, it also offers more possibilities for SCM integration with external vendors and partners. SCM could help libraries proactively address previously mentioned challenges in serving patrons by building a much more flexible network of operations that mitigates risks and supports innovations. As libraries embrace team and collaborative cultures that cross groups and functional departments, they might consider SCM to be a priority for their institutions.

One point to clarify is SCM's role in contributing to the financial success of an organization. Financial success should not be measured by the bottom line for libraries, but rather from the perspective of financial stewardship. Libraries are in the business of stewarding resources to build excellent services. Financial stewardship speaks to the utilization of funds for maximizing services. In this sense, SCM is the perfect tool to meet expectations of the stakeholders and to differentiate products between libraries and other information providers.

CASE STUDIES RELEVANT TO LIBRARY BUSINESS

Coyle et al. (2013) described five utilities provided by a business: form, possession, place, time, and quantity. The form utility determines the finished products. In the library system, the finished products extensively come from suppliers. Vendors have mostly predetermined the characteristics and features of those final products. Whether it is a print book or an e-book, the weight and dimension of the book or the downloadable chapters and integrated citation services of the e-book is set for libraries. In other words, libraries have no control over the form utility of their collections. Possession utility is the work of the marketing process of business; creating desires and wants, facilitating sales, and transferring ownership of the products. Libraries are in the business of collection loans. There is no transfer of property per se, but marketing can indeed create value in lending services and improve patrons' awareness of library services.

Place, time, and quantity utilities for products are the missions of SCM. Place utility is about moving products from production points to market locations and making sure the products and services are readily available in places for patrons. Place utility is the core service that libraries offer to their patrons. Libraries must master this utility.

Time utility is the availability of products when users need them. SCM coordinates the movement of inventories from production and storage to users. In SCM, just-in-time delivery of parts or merchandise is an essential component. In libraries, just-in-time has taken precedence and gained more popularity in the past 10 years. Demand-Driven Acquisition or Patron-Driven Acquisition is a recent manifesto of the time utility. The difference is that libraries leveraged just-in-time to control spending, but SCM drives up value by providing immediate access to products. Time utility will be one of the main topics of this book. It will help libraries meet user requirements and satisfy their needs.

Quantity utility ensures availability of the right amount of products. Too much or too little stock is either raising costs or losing revenue. SCM applies strategies in forecasting, scheduling, and inventory to level quantity according to need. These strategies may seem irrelevant to libraries, but an approval plan is a form of forecasting the needs of users. The management of approval plans has a direct impact on users. Even so, quantity utility for libraries is unique in comparison with other industries.

It is important to note here that place, time, and quantity all play a part in producing value to patrons. SCM also concerns itself with variety, condition, and price of products, and these points are on the minds of library professionals, too. Supply Chains must deliver the Seven Rights of Fulfillment: providing the right product, to the right customer, at the right time, at the right place, in the right condition, in the right quantity, at the right cost (CSCPM et al., 2013). These seven rights will serve patrons better, faster, and cheaper, which is a puzzle that all libraries need to solve in the current economy and social context. It is imperative for libraries to not only satisfy customer needs but also do so as efficiently as possible.

This section gives one example from the retail industry, Amazon, and one from the service sector, Netflix, based on the nature of library business:

- Libraries' vendors predetermine the form utility of library products.
- Libraries have preestablished complex processes and have built facilities to provide library services.
- Libraries are in the middle of the Extended Supply Chain of the knowledge and information industry.
- Libraries have been moving up the Chain lately by taking on the knowledge creation role.
- Libraries need to deliver services both physically and virtually within their physical and online spaces.

Although the two case studies are closely related to the types of products that libraries offer, they do not precisely mirror every service of libraries. These models of SCM integration work for the retail and service sectors, particularly these two innovative firms. They can contextualize core ideas of SCM and contribute some food for thought about how the library could work across departments, peer institutions, and its suppliers to maximize value for users.

Amazon's Supply Chain shares many similarities with libraries. Although Amazon expanded to an online department store from an online bookstore, part of its business is still in print and digital publications. Although libraries are not in retail, Amazon's example can still offer ideas about how to reconstruct collection processes.

SCM has been central to Amazon's online retail experience since the beginning of the firm in 1995, and it is also a vital strategy that gives Amazon an edge over other competitors. Amazon has been improving SCM from the angle of the Extended Supply Chain. In 2015, it topped Gartner Inc.'s Top 25 Supply Chains because of its constant innovation and a consumer-centric approach that emphasizes frugality and works from the customer backward (Fraser, 2016).

Amazon concentrates on a superior and effortless experience (Schuman, 2016). Amazon developed strategies to increase place, time, and quantity utilities of its products. Amazon owns facilities, such as fulfillment centers, distribution center networks, sorting centers, and also outsources to third-party facilities (MWPVL, n.d.). Amazon stores only limited inventory for the most needed items and leverages sellers' storage for the rest. For example, Prime Now Hub only holds the highest velocity items such as best-seller books. Serguei Netessine, professor of operations and information management at Wharton, states "Amazon keeps the most popular products in inventory, but uses a mix of techniques to deliver goods." Using drop shipping, Amazon also has real-time links to manufacturers, who ship goods directly to consumers on its behalf. Kindle—the company's much talked about e-book reader—is an example of this. Customers around the world can use 3G wireless technologies for faster downloading, among other things (Wharton University of Pennsylvania, 2009). Amazon has advanced systems and tools for real-time data from which they can gauge demands and manage stocks accordingly. Amazon is transforming itself into a logistics company, as it facilitates the storage and sale of goods from third-party vendors (Fraser, 2016).

For years, Amazon has invested heavily in reducing the delivery time of its products. According to Soper (2016), Amazon recently embarked on an aggressive global expansion of its fulfillment service, Dragon Boat. The project will tighten direct relations between many manufacturers and Amazon patrons around the globe by creating a global delivery network that controls the flow of goods from factories in China and India to customer doorsteps in Atlanta, New York, and London. Amazon will expand to shipping container wholesaling, entirely bypassing transit and cargo brokers. Not only does the project offer reduced rates for space on trucks, planes, and ships, but it also means that Amazon amasses inventory from thousands of merchants around the world. Amazon described "Dragon Boat" as a "revolutionary system that will automate the entire international supply chain and eliminate much of the legacy waste associated with document handling and freight booking." Schuman (2016) indicated, "a planned global supply chain move by Amazon could disrupt product access, sharply lower Amazon's costs and accelerate product delivery to shoppers."

Amazon's endeavors were not about trying to earn more profits but rather to increase its competitiveness and to gain market share. According to Schuman (2016), Jeff Bezos, CEO of Amazon, will forego profits every time if it means boosting market share and eliminating competition. Dragon Boat was Amazon's rather sophisticated twist on Walmart's original attack plan for smaller physical chains. Needless to say, by doing so Amazon improved its profitability; however, foremost was the gain of customer loyalty through their efforts to create a better shopping experience in terms of product cost, variety, availability, fulfillment, delivery, and postsale service. In the author's opinion, libraries are facing similar challenges in "market share" and "competition."

Netflix started as a DVD rental by mail business in 1998. As the firm grew, it transformed into an online video streaming service in 2007 and eventually started to expand into the "content producing" business in the past few years by investing in original and exclusive content, such as "House of Cards" and "Orange is the New Black." Regardless of some mishaps in the timing of the decision to switch from DVD rental to an online streaming enterprise, Netflix has been incredibly successful in its 20-year history. By 2016, Netflix expanded their services to over 130 countries (Impiger Technologies, 2016).

During its early years, what set Netflix apart was not their DVD collection, but customer service and getting those DVDs turned around quickly (Sandoval, 2012). This customer-centered philosophy extended to their digital domain of work. Netflix understood the values that the entire supply chain offered to achieve user satisfaction. Kevin McEntee, VP Digital Supply Chain, states "The content owning studios need to innovate on the supply chain so that they can develop an effective, permanent, and growing sales channel for digital distribution customers like Netflix. Finally, post-production houses have a fantastic opportunity to pivot their businesses to eliminate this complexity for their content owning customers" (2012).

McEntee went on with his vision of the Netflix Digital Supply Chain, "the Digital Supply Chain is an online experience for content buyers where they browse available studio content catalogs and make selections for content to license on behalf of their consumers. Once an 'order' is completed on this system, the materials (video, audio, timed text, artwork, meta-data) flow into retailers systems automatically and out to customers in a short and predictable amount of time, 99% of the time. Eliminating today's Supply Chain complexity will allow all of us to focus on continuing to innovate with production teams to bring amazing new experiences like 3D, 4K video, and many innovations not yet invented to our customer's homes." This vision calls for transformation of content acquisition, inventory, and delivery. Netflix planned to reengineer on two critical components: (1). content quality, video, audio, text, artwork, and descriptive metadata for all of the needed spoken languages, and (2). processes and tools that enable business-to-business ordering and catalog management, enabling global online systems to track content orders, and to curate content records. Back in 2012, Netflix had a set of tools for managing content orders and curating its content catalogs. They are internal, and Netflix worked with the industry via phone calls and emails containing spreadsheets of content data. Netflix then released a plan to build a Content Partner Portal, in which delivery of data and real-time feedback on performances (timeliness and error rates) can be shared regularly and in real time. The portal not only will help content owners to manage efficiency of deliveries but will also give Netflix the ability to objectively assess content owners when planning pricing for content. Netflix also created a new group, Content Partner Operations, which went upstream to help postproduction houses and content owners prepare content for Netflix. It acts as an advocate for Netflix Delivery Specification, which gained support throughout 2013 from video and audio technology partners. Additionally, the Netflix certification process already resides in postproduction houses to streamline onboarding new content owners and dissemination of films around the world. Impiger (2016) pointed out "the amazing supply chain management, which is essentially the core pillar of Netflix business."

REFERENCES

Ayers, J. B. (Ed.). (2000). *Handbook of supply chain management.* Boca Raton, Florida: CRC Press.

Blanchard, D. (2010). *Supply chain management best practices* (2nd ed.). Hoboken, New Jersey: Wiley.

Coyle, J. J., Langley, J., Novack, R. A., & Gibson, B. (2013). *Supply chain management: A logistics perspective.* Mason, OH: Cengage Learning.

CSCPM, Chen, H., Defee, C. C., Gibson, B. J., & Hanna, J. B. (2013). *Definitive guide to integrated supply chain management: Optimize the interaction between supply chain processes, tools, and technologies* (1st ed.). Upper Saddle River, New Jersey: Pearson FT Press.

Ellram, L., & Cooper, M. (1993). Characteristics of supply chain management and the implications for purchasing and logistics strategy. *International Journal of Logistics Management, 4*(2), 1–10.

Fraser, A. (June 23, 2016). *Amazon's supply chain process: How do Amazon create one of the world's best supply chains?* Innovation Enterprise. Retrieved from: https://channels.theinnovationenterprise.com/articles/amazon-s-supply-chain-process.

Impiger Technologies. (January 8, 2016, Impiger Technologies, Retrieved from:). *Supply chain management lessons from Netflix.* https://www.impigertech.com/blog/supply-chain-management-lessons-from-netflix/.

Lummus, R. R., & Alber, K. L. (1997). *Supply chain management: Balancing the supply chain with customer demand.* Falls Church, VA: APICS Educational & Research Foundation.

Lummus, R. R., & Vokurka, R. J. (1999). Defining supply chain management: A historical perspective and practical guidelines. *Industrial Management and Data Systems, 99*(1), 11–17.

Monczka, R. M., & Morgan, J. (1997). What's wrong with supply chain management. *Purchasing, 122*(1), 69–73.

MWPVL. (n.d.). Amazon global fulfillment network, MWPVL International, Retrieved from : http://www.mwpvl.com/html/amazon_com.html.

Oxford University Press. (n.d.). Supply chain. Oxford English Dictionary. Retrieved from: http://www.oed.com/view/Entry/194665?redirectedFrom=supply+chain#eid.19718112.

Quinn, F. J. (1997). What's the buzz. *Logistics Management, 36*(2), 43–47.

Robinson, A. (January 23, 2015). *[INFOGRAPHIC] The evolution and history of supply chain management.* Cerasis. Retrieved from: http://cerasis.com/2015/01/23/history-of-supply-chain-management/.

Sandoval, G. (July 11, 2012). *Netflix's lost year: The inside story of the price-hike train wreck.* Cnet. Retrieved from: https://www.cnet.com/news/netflixs-lost-year-the-inside-story-of-the-price-hike-train-wreck/.

Schuman, E. (February 12, 2016). *What Amazon is doing with its supply chain could devastate the competition.* Computer World. Retrieved from: http://www.computerworld.com/article/3032656/retail-it/amazons-supply-chain-move-could-prove-devastating.html.

Soper, S. (February 9, 2016). *Amazon building global delivery business to take on Alibaba.* Bloomberg Technology. Retrieved from: http://www.bloomberg.com/news/articles/2016-02-09/amazon-is-building-global-delivery-business-to-take-on-alibaba-ikfhpyes.

Supply Chain & Logistics Institute. (2015). *Evolution of SCL.* Georgia Tech Supply Chain & Logistics Institute. Retrieved from: https://www.scl.gatech.edu/about/scl/history.

Sweeney, E. (2011). Supply chain integration: Challenges and solutions. In *Information Resources Management Association of United States of America, supply chain management: Concepts, methodologies, tools, and application* (pp. 221–246). Hershey, PA: Business Science Reference, IGI Global.

Wharton University of Pennsylvania. (November 11, 2009). *Fit for the holidays: Amazon is shaping up and shipping out.* Knowledge@Wharton. Retrieved from: http://knowledge.wharton.upenn.edu/article/fit-for-the-holidays-amazon-is-shaping-up-and-shipping-out/.

Wikipedia. (July 7, 2016). *Supply chain.* Wikipedia. Retrieved from: https://en.wikipedia.org/wiki/Supply_chain.

FURTHER READING

Davis, D. (1995). State of new art: Manufacturers and trading partners learn as they go. *Manufacturing Systems, 13*(8), 2–10.

Dittmann, P. J. (2012). *Supply chain transformation: Building and executing an integrated supply chain strategy.* New York City: McGraw-Hill Education.

PART 2

Library Supply Chain Model

CHAPTER 4

Supply Chain Strategies for Libraries

A Supply Chain, boiled down to its essential elements, is the sequence of events and processes that take a product from dirt to dirt (Blanchard, 2010). Any discussion of the Supply Chain can legitimately be broad or narrow depending on the perspective of the "definer" and the interests of those involved in the conversation (Ayers, 2000). Chapter 3 explains that supply chain definitions vary based on different industries. It is nearly impossible to have a one-size-fits-all supply chain model for all sectors.

The notion of"the Extended Supply Chain" (Fig. 4.1) should encourage any organization to consider the contextual environment of their supply chain, which includes their vendors' suppliers and the customers' clients. Although the positioning of a particular institution on the chain is unique, depending on how close it is near the raw material or near the products, the extended view is a way to engineering its supply chain without losing the bigger picture. The reason for having a contextual view is simple: unless the chain was designed holistically across all entities of the extended chain, its effectiveness and its return on value (could be interpreted in dollars or user impact) would be significantly limited. The above statement does not mean that design or improvement of a supply chain has to start with meeting all key personnel of each entity; though it might be a perfect approach, it is arduous to achieve in the current environment. Improvement of the supply chain may start internally across the organization by breaking down the silos. Poirier (1998) notes that moving a supply chain optimization from an internal to an external focus is an especially significant barrier, as the systems and people are just not in place and the required behavior is counterintuitive.

Supply Chain must start with customers, not assumptions of what customers need. What an organization learns from customers will build a solid foundation for supply chain improvement. Another fundamental concept of Chapter 3 is the bidirectional nature of Supply Chain Management (SCM). Knowledge of user requirements should be shared upstream with vendors and their vendors, which will bring entities together solving Supply Chain issues to improve service quality.

Supply Chain Management for Collection Services of Academic Libraries
ISBN 978-0-08-102031-9
http://dx.doi.org/10.1016/B978-0-08-102031-9.00004-9

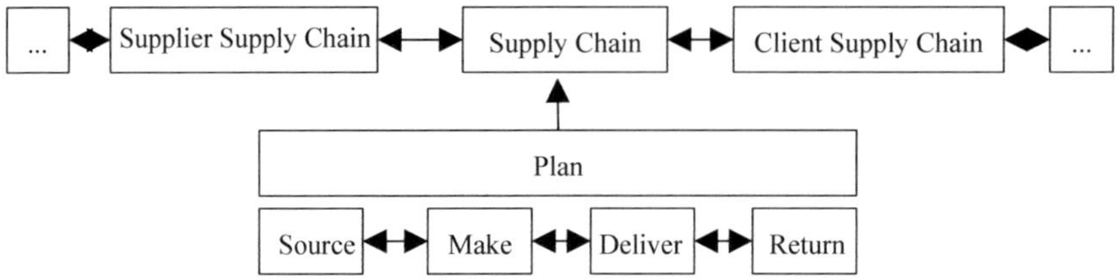

Figure 4.1 Extended supply chain.

Libraries are in a unique position. They are closer to information consumers on the extended chain of the knowledge industry. What libraries learn from their users can help vendors meet libraries' mission and therefore serve patrons. Such a relationship and knowledge sharing exist between libraries and vendors at the moment; however, it is often within the boundary of products, and it is used as a means to improve existing ones. Perhaps libraries could take a leadership role to formalize a new dialog with their vendors around the topic of supply chain improvement and innovation rather than solely focusing on particular products. Ayers (2000) indicated that supply chain thinking has untapped potential for maintaining a competitive edge or moving a company from an unfavorable to a more advantageous position. Supply Chain studies the flow of resources, processes, finances, and knowledge. Product feedback improves the product, but Supply Chain conversations may produce different products, solutions, and results that are much more streamlined to serve users.

COLLECTION SERVICES LIFE CYCLES MODEL

To ensure a proper design of a Supply Chain that leads to a successful implementation, a business needs to determine the type of its operations. There is plenty of research helping firms to figure out the DNA of businesses. The most important one is to understand the current position of a firm in a market, particularly, the products or services in the current market. One of the models is the Product Life Cycle Model created by Ayers (1999). The product life cycle model establishes four states of any product or service, which are inception, growth, maturity, and decline.

A business needs to honestly assess its position and the state of its products, which may be in different stages. For instance, libraries have experienced decline in reference services, but a surge in research management services. The focus of this book is collection services; however, it might additionally be valuable to go through a similar exercise to determine the state of other library products.

There are many nuances about collection services. Many librarians understand the current library market and know that circulation of physical items has been going down for majority of Northern American academic libraries, while direct access to electronic resources has been on the rise. The following is the author's personal analysis of library collection services, which others may see differently. The purpose of the author's thought exercise is to encourage conversations among library associations, librarians, and other information professionals, via which to drill deeper to the current situation of collection services, and therefore map future strategies for enhancing the values of collection services to library patrons.

Print Collections

No doubt that the majority of libraries have been experiencing a decline in print circulations since the advent of digital collections. Anecdotes, and literature (Haley & Chu, 2010; Rose-Wiles, 2013), have confirmed the decrease. Overall, print collections are in the declining phase of their life cycle. Even for disciplines heavily dependent on print, one can safely say that the circulation is not going up and usage remains stagnant. Therefore, one could deduce that even for those academic disciplines, print arrived at a maturity state, according to the Ayers Model.

The only exception is special collections vis-à-vis general collections. As both categories support research, teaching, and learning needs, special collections are truly collectibles as their values increase over time. They are assets, in the author's point of view, compared with general collections. Common collections are available and accessible through many information providers; they are much more likely duplicated across libraries. Although the current school of thought is to recognize them as assets, it is arguable whether they shouldn't be considered liabilities. If they are not used or circulated, and they take much storage and resources to acquire and maintain, their value to institutions is questionable. Considering all the investments, such as acquisition funds, facility costs, equipment, cooling and heating, and labor, libraries need to figure out a way to calculate the values of their general collections to justify the investments before their stakeholders.

The author here is not trying to undervalue general collections or print collections. If collections are services that libraries provide to support their faculty and students, library professionals need to consider their return on value to patrons. Libraries must create strategies to maximize the collection benefits to users. Just as the Part one—Introduction to Supply Chain Management and Case Studies of the book indicated that libraries are extremely underresourced, the understanding of the market positions of collection services will pay dividends in realizing library missions and increasing library relevance to both parenting organizations and patrons.

For services in the Maturity state, the Life Cycle Model recommended supply chain as a way to move the product into other states of the lifecycle. Ayers (2000) described Maturity as "Demand stops growing. Supply chain features dominate over product features. Product features are common among competitors. Niche strategies appear, with competitors focusing on particular market segments, using supply chain design as a strategy." Ayers' statement depicts the current state of general print collections accurately. The high-demand days of those collections are plateaued and even falling. Features or functionalities of the print are identical among many providers; and many libraries hold duplicates. The only advantage that libraries have over other providers, such as bookstores, is perhaps the "free" nature of their collections. And even then, other merchants may offer larger inventories, more variety of selections, instant access, and quicker delivery. All the above factors affect a patron's information-seeking behaviors. Segments of library demographics and nonlibrary information providers were never mutually exclusive; however, lately library patrons have more incentives to tilt toward other vendors than ever before because more choices exist. Ayers Model gave libraries a road map to improve their supply chain by innovating it with features of the extended product.

How long print as a predominant format would stay in its maturity state is difficult to predict. Libraries should improve supply chain by increasing the value of their print collections to users; at the same time, libraries must also prepare for the next stage of their print collection services, the Decline stage. The trend has started in some disciplines already, and it will become more widespread in the future. Once a product moves to the Decline phase, it is impossible to reverse the cycle by moving it back to its prime time. The only option is to innovate by transforming the product. For example, electronic medium has been emerging and will continue to make its way into the mainstream formats, although they are still considered to be in their Inception or Growth stages compared with print.

Electronic Collections

Ayers (1999) described products in Inception phase as follows, "Producers compete in a start-up mode. The products are new and markets are forming. Supply chains are taking shape. No dominant supply chain model has emerged." Electronic collections entered libraries more than two decades ago, but it is also hard to say if electronic resources have passed their inception state yet. Publishers, aggregators, libraries, and library patrons are still trying to figure out how electronic resources impact their bottom lines. Regarding offerings, technologies, and standards of e-resources, there is not a clear winner that has emerged from competitions at this moment. Licensing terms, purchasing options, system environments, and access policies of digital resources continue to vary among vendors.

At the same time, one may argue that digital collections are in the growth phase rather than in the inception phase. Certain fields, such as engineering and sciences, have eagerly adopted the electronic format. For example, high-energy physics are extensively using online open access services, such as arXiv. For majority of library patrons, electronic collections are also gaining popularity. The only thing that is holding their further adoption, in the author's point of view, is usability issues. Libraries and library patrons are interacting with proprietary digital technologies, as publishers are looking for new ways to package e-resources and regulate access and use of their e-publications.

Digital content is also growing based on patrons' requirements. Acquisitions of born-digital materials, such as research data packages and digital archives, are becoming a norm. No matter which of the two states electronic collections are in, Inception or Growth, e-resources are going to stay and expand.

Supply Chain is an obvious winning strategy that will give libraries competitive advantages. Ayers (1999) believed "once products or services moved to their mature phase, winners have been sorted out based primarily on product, and, to a lesser extent, on supply chain features." Libraries should anticipate the future of electronic collections and create appropriate supply chain strategies that will maximize the impacts of their electronic resources on user goals.

SUPPLY CHAIN SEGMENTATION AND ROADMAP MODEL

Perez (2013) introduced six generic supply chain types: efficient, fast, continuous-flow, agile, custom-configured, and flexible. His work was based on Marshall Fisher's revolutionary concept of supply chain segmentation in 1997 and several models developed by other academics and consultants following Fisher. Perez's model fixed some major shortcomings of previous models that overlooked the combinations and connections between the major drivers throughout the value chain, the alignment with industry's competitive framework, and with each organization's unique value proposition.

Perez developed the "Supply Chain Roadmap" model by modifying previous models. It is based on the understanding of the industry, unique values of an organization within the industry, and competitions, the organization needs to create management focus and improve its internal processes. The six models are categorized into two groups: supply chain models that are oriented toward efficiency and include efficient, fast, and continuous-flow supply chains and those that are oriented toward responsiveness and include agile, custom-configured, and flexible supply chains.

To understand the type of business that libraries are, the first step ought to be pinpointing the right supply chain models for their products, therefore choosing the right management foci and strategies to improve supply chain processes. Libraries offer a variety of services. Library space, equipment, and facility are a type of service; discovery is another kind; collections is the apparent core service; and other derivative services from collections are circulation, reference, consultation, preservation, conservation, document delivery, and interlibrary loan. Each of the above services resides at different life cycles and probably requires different strategies for improvement, transformation, or, in some cases, even cessation. The models that the author recommended, the Supply Chain Operations Reference (SCOR) Model and the Supply Chain Roadmap model, which will be discussed in later chapters, will offer two excellent frameworks for libraries to start those strategic conversations.

The author emphasizes the need for applying the two Models to collection services. Libraries differentiate special collections, general collections, teaching collections, research collections, and reference collections. So do their supply chain strategies. Taking special collections as an example, one may further break them down into rare books collections, archival collections, manuscript collections, etc. Each of the types can be on its own distinct life cycle, and hence needs different supply chain processes and management focus. The author, based on his understanding of the library operations, recommends libraries adopt and greatly customize the two models of the six generic models to fit in library-unique environments and the knowledge/ information industry.

The "Efficient Supply Chain" Model

The Efficient Supply Chain is one of the three models oriented toward efficiency along with fast and continuous-flow models. This model is best suited for industries that gear toward low-cost and/or high relevancy of asset utilization to total cost (Perez, 2013). The consideration for adoption of such a model is to meet accountability expectations of library stakeholders and to better steward library resources in meeting the increasing demands from the academic users under current difficult financial times.

Most importantly, speaking from the context of knowledge access and consumption, libraries are facing intense competition with other information providers. For example, search engines and online social sites are replacing the library reference service. In some cases, information consumers do not see significant differentiations between libraries and other providers. The unique values of the libraries are difficult to recognize by the untrained eye. The "Efficient" Supply Chain Model was designed for such an environment.

Before libraries consider implementation, it is necessary to modify the Model, as it is set for industries competing most solely for decreased price of their product or service. As libraries are also competing for the preferred mode of access and other factors, the adapted "Efficient Model" should take into account the above factors as well. It is also true that libraries have an edge on price over other information providers because the services are free of charge. Because the services are free to users, majority of them take services libraries provide for granted. Not all individual professionals understand the strategic coherence of the acquisitions budget as a whole; rather, they see how it relates to their individual narrow work needs. Such siloed view on collection services contributes to the inefficient utility of library resources. The cost side of considerations necessitates the needs to reengineer the "Efficient" Model based on libraries' unique situations and cultures. In later chapters, the author will drill down to the details of the Model and suggest strategies that libraries may adopt to improve their supply chain of the collection services.

The "Flexible" Supply Chain Model

The "Flexible" Supply Chain is one of the three models oriented toward responsiveness along with agile and custom-configured models. This model is best suited for companies that must meet unexpected demand and therefore are faced with high-demand peaks and long periods of low workload (Perez, 2013). The rationale for adopting such a model in libraries is to respond to the changing users' information-seeking behaviors and to maximize positive impacts of libraries on their patrons. This model is centered on adaptability, the ability to reconfigure internal processes to meet a customer's specific need or solve a patron's problem (Perez, 2013). Most of the library professionals can resonate with unique needs of individual patrons and particularly how their needs have evolved rapidly in the last decade.

Patrons, on the other hand, greatly appreciate the libraries' ability to solve their problems quickly. This model provides strategies to offer much-needed flexibility for libraries to deal with "unexpected," "extraordinary," "emergent," and "above-and-beyond" scenarios.

Again, modifications to the model are needed for library operations. This model is produced for industries with unpredictable demands and where business is highly regulated by market median price. With libraries responding to the expansion of the research reach of the universities in the past 5–10 years, unpredictability is a new norm to predict new collection demands. As research practices are still evolving, libraries have to be able to adapt to changes quickly. Research collection curation is entirely different from what libraries know about collection services. The turnaround time, technical and technology support, and infrastructure dependence are extremely critical, more so than other collections libraries typically collect ever were. Again, libraries are not a business that is evaluated by market prices, so this model has to be altered with library-unique value propositions to end users in mind. In later chapters, the author will elaborate on those changes to the "Responsive" Model and recommend strategies that may improve libraries' collection supply chain.

Other Supply Chain Models

Author will not introduce the Fast, Continuous-Flow, Agile, and Custom-Configured Models not because they are irrelevant to the libraries in general, but because this book's focus is on Collection Services. Those models might be highly applicable for other types of library services and should be explored separately. There is value for library professionals to understand those models; however, they won't be the foci of this book. Later chapters of the book will develop examples of the collection services supply chains based on the characteristics of the Efficient and Flexible Models. At this point, readers of the book might question how libraries may develop nonconflicting strategies and practices with two seemingly contradictory models. Acknowledging nuances among different types of collections, libraries have to create different supply chains for each. With that said, the author believes that the two proposed models for libraries can be complementary depending on the types of collections. After all, libraries should design supply chains that address both efficiency and flexibility. The ultimate goal of any supply chain is to meet user requirements, and both these models aim at delivering target responsiveness to user needs. Later chapters will look at how to combine the two Models.

SUPPLY CHAIN OPERATIONS REFERENCE MODEL

SCOR (APICS, 2015) defines SCM as plan, source, make, deliver, and return (see Fig. 4.1). This view offers a framework in which any industry may contextualize and model a supply chain. As we understand how actions to plan, source, make, deliver, and return translate into distinctive processes for individuals, organizations, and industries, the SCOR Model serves as a tool for facilitating communication and conversations and a starting point for many institutions to design their supply chains and management. As this book is written for library professionals, regardless of their roles, such as collection development, IT, technical services, or public services, SCOR creates a boundary for library workers to structure dialogs on SCM for their own libraries and the knowledge management field. The next chapter will provide emphasis on the SCOR Model.

REFERENCES

APICS. (2015). *SCOR framework*. APICS Supply Chain Council. Retrieved from: http://www.apics.org/sites/apics-supply-chain-council/frameworks/scor.

Ayers, J. B. (1999). Supply chain strategies. *Information Strategy: The Executive's Journal, 15*(2), 3–10.

Ayers, J. B. (Ed.). (2000). *Handbook of supply chain management*. Boca Raton, Florida: CRC Press.

Blanchard, D. (2010). *Supply chain management best practices* (2nd ed.). Hoboken, New Jersey: Wiley.

Haley, D., & Chu, C. M. (2010). *The use of print materials in the internet age: A comparative study of academic library circulation patterns* (ProQuest dissertations and theses).

Perez, H. D. (March 06, 2013). *Supply chain strategies: Which one hits the mark? (CSCMP and supply chain media, LLC)*. Supply Chain Quarterly. Retrieved from: http://www.supplychainquarterly.com/topics/Strategy/20130306-supply-chain-strategies-which-one-hits-the-mark/#fn7.

Poirier, C. C. (1998). The path to supply chain leadership. *Supply Chain Management Review, 2*(3), 16–26.

Rose-Wiles, L. (2013). Are print books dead? An investigation of book circulation at a mid-sized academic library. *Technical Services Quarterly, 30*(2), 129–152.

FURTHER READING

Hudson, S., & SCRC (October 27, 2014). *The SCOR model for supply chain strategic decisions*. NC State Poole College of Management Supply Chain Resources Cooperative. Retrieved from: https://scm.ncsu.edu/scm-articles/article/the-scor-model-for-supply-chain-strategic-decisions.

CHAPTER 5

Supply Chain Model for Libraries

SCOR MODEL, VALUE CHAIN MODEL, AND LIBRARY SUPPLY CHAIN LITERATURE

This chapter concentrates on a possible library Supply Chain of Collection Services based on the Supply Chain Operations Reference (SCOR) Model. The goal of the chapter is to set up a framework or a structure that libraries may have conversations around and that may help explain the concept to library professionals.

The SCOR Model was described in Chapter 4. It suggests a flow of activities or processes that get products and services into the hands of users. The activities should include informational, financial, decision making, and clear material flow, and the flow should not be assumed to be one directional. Rather, the flow should go two ways, downstream and upstream, and should better start backward with defining customer needs first. In a nutshell, SCOR activities' flow describes a series of stages to plan, source, make, deliver, and return. The chapter will mirror those stages with the library collection workflow in mind and include operations of collection planning, acquisitions and licensing, collection enhancements, delivery to stacks or document delivery, and returns.

Supply Chain Management (SCM) interacts with another critical model, Michael Porter's Value Chain Model. Porter's model deals with the added value of the flow of activities aimed at getting a product or service to users. The author wrote an article on Research value chain and its relationship to the library value chain based on Porter's model, in which Porter's original model was introduced (Wang, 2013):

> *Michael Porter (1985) introduced the value chain concept in his book Competitive Advantage: Creating and Sustaining Superior Performance. The value chain analyzes activities of an organization and relates them to its competitive position in the market. In summary, the value chain disaggregates an organization's operations into a chain of primary activities and a set of support activities. The chain of the primary activities is the flow of the core steps via which a product (or a service) is produced. Support activities enable primary activities. Costs will increase with each primary activity and value should increase with each primary activity. The sum of value should be greater than the sum of the costs. More importantly, the values are what customers are willing to pay for and are competitive to the product offerings*

Supply Chain Management for Collection Services of Academic Libraries
ISBN 978-0-08-102031-9
http://dx.doi.org/10.1016/B978-0-08-102031-9.00005-0

of its competitors. If an organization seized to create value for end users or if the added value was not comparable to those of its competitors, it would lose competitiveness. A value chain is a framework by which the organization evaluates its operations and thus creates strategies to increase its competitiveness. The original value chain was modeled for for-profit organizations; however, the essence of the concept can be generalized to many other sectors. Since both libraries and researchers perform a series of activities to "manufacture products," the author will apply the value chain to faculty and students research activities as well as the library activities to deliver more relevant arguments and analysis in this paper.

When modeling library operations with the value chain, libraries will need to answer two fundamental questions:

1. Do the primary activities still provide value to today's users?
2. Does the value created through those primary activities outweigh the costs to produce library services?

When libraries attempt to model its supply chain, those are important questions to ask. The value must be created for users in a unique way from that of other providers; additionally, value determination must be based on the decision of whether to keep or eliminate certain procedures, tasks, or workflows. The Value Chain is a way to ensure that what an organization implements is strategically aligned with its users in the same way the supply chain aligns with user requirements. Without the Value Chain Model, libraries may design a perfect supply chain of little meaning; without the Supply Chain Model, libraries waste resources and lose the opportunities to run their business efficiently.

The Supply Chain is a method of bringing competitive advantages to an institution. In his work, Porter also acknowledged the importance of supply chain in its connection and bonding to individual activities and functions throughout the Value Chain so as to address the unique business value proposal to customers (Porter, 1980). Perez (2013) paraphrased Michael Porter, "While operational efficacy deals with achieving excellence in individual activities or functions, supply chain strategy defines the connection and combination of activities and functions throughout the Value Chain, to fulfill the business value proposal to customers in a marketplace." So the most important issue is value creation. SCM goes beyond individual and functional operational efficiency.

Based on the article by Kress and Wisner (2012) and literature scan, current application of the Supply Chain concept is rather limited in the library industry. Kress and Wisner applied the Supply Chain modeling to a merger between two departments, User Services and Technical Services. Both of them mapped an information resources supply chain for an academic library, aligning processes and establishing performance measures. It was evident from their research that Supply Chain, the term, has been referenced since 1996 by Cornish to advocate collaboration between intellectual property owners and libraries to improve information flow to meet the needs of users. Ball and Wright (2000) even described the role of libraries as intermediaries along the primary library supply chain processes of creation, publication, aggregation, access, and use. Back in 2000, Ball and Wright saw libraries as owners of aggregation, access, and use processes, which remain true today; however, since then, libraries have also been taking on roles of creation and publication of content. Meng-Xing, Chun-Xiao, and Yong (2010) diagramed a digital libraries supply chain, composed of content suppliers, digital library alliances, content service providers, third-party service contractors, and, finally, the library users, which later chapters will reference. Discussions and empirical analyses of library supply chains remain quite limited, specifically on collection services. One major detail that the author calls out is that the previous supply chain models are universal models designed for the entire industry, which libraries are part of. Although those models promote the fundamental value of the supply chain, satisfaction of end-user goals, collaborations along the chain, and process and workflow improvement, those models were not designed specifically for libraries.

LIBRARY COLLECTION SUPPLY CHAIN MODEL

This chapter will focus on developing a Collection Supply Chain specifically designed for libraries. The model will promote all supply chain values, but more so, it will emphasize the improvement and optimization of specific collection processes. The Collection Supply Chain is guided by the SCOR Model.

Huan, Sheoran, and Wang (2004) classified SCM research into three categories: operational, design, and strategic. Depending on where an industry is on the spectrum of adopting and applying SCM, each type may bring a different level of impact. Generally speaking, with the current development of SCM, many analytical and numerical models have been proposed to handle supply chain operations and design issues (Chopra and Meindl, 2001). Operational issues are often concerned with daily functioning of a facility. Design issues focus on the location of decision spots and the objectives of the chain (Mourits and Evers, 1995). Design models are tactic approaches that strive toward internal integration and optimization of the chain rather than those of the extended chain.

Given that SCM is a relatively new concept for the library industry, libraries can most benefit from the principles of strategic models. Strategic models were scarce, and the most promising model is the SCOR Model (Huan et al., 2004):

- Standard descriptions of management processes;
- A framework of relationships among the standard processes;
- Standard metrics to measure process performance;
- Management practices that produce best-in-class performance;
- And, standard alignment to software features and functionality.

The discussions, guided by the SCOR Model, hopefully will expose some of the profession's deeply rooted assumptions about the current collection processes. Once libraries establish new process strategies based on the Model, libraries may further the adoption of SCM by applying other models that are dealing with design and operational issues.

SUPPLY CHAIN OPERATIONS REFERENCE MODEL

The SCOR Model was created in 1995 and has been refined ever since (Blanchard, 2010). Supply chain Council published the SCOR model, which is the world's leading supply chain framework, linking business processes, performance metrics, and practices and people skills into a unified structure (APICS, 2015; Hudson and SCRC, 2014). SCOR aims to increase the speed of system implementations, support organizational learning goals, and improve inventory turnover (APICS, 2015). The newer version published in 2015 also includes an element of "enable," which will be explained in Chapter 16. For now, the introduction of the Model in this chapter is an attempt to create a guide that focuses on flows of products and services and decision making on finance, resources, and processes. SCOR top level processes include the following:

Plan

First of all, Plan means to balance demand and supply. Plan also includes a firm's understanding of its available resources and constraints, and how those available assets or limitations support or limit the management of demand and supply. Strategies and initiatives should be set up, and so should the guidelines, rules, and changes that span key components, such as inventory, transportation, assets, and regulatory compliances. Finally, it is critical to determine communication and feedback mechanisms on performances of new strategies and initiatives along the entire chain, both downstream and upstream.

Source

Source involves all the infrastructure, processes, systems, and evaluation mechanisms to acquire necessary materials to meet planned demands. APICS describes Source as managing inventory, the vendor network, supplier agreements, and supplier performance. It discusses how to handle vendor payments and when to receive, verify, and transfer product.

Make

Make not only includes the production of the final products or services but also includes processes of packaging, staging, and releasing them. This grouping then has to include production network, equipment, facility, and transportation.

Deliver

Deliver encapsulates all the activities that transport the finished goods to consumers. It typically includes customer ordering, invoicing, warehousing and life cycle management, and transportation (could be both international or domestic).

Return

Return is part of the customer service support. It involves activities that transport the products back from customers. It depends on policies, facility, and transportation. It can be viewed as a reverse of the Deliver processes in the library setting.

Enable

This is a new process that was introduced in Version 11 of the SCOR Model (December 2012), where the Supply Chain Council gives a framework for assessing Supply Chain and recommends a multilevel process to manage Supply Chain projects. SCOR Model combines business process reengineering with benchmarking, best practices, and process measure into an all-encompassing framework for executing a supply chain (Blanchard, 2010).

PLAN—COLLECTION PLANNING

In the SCOR Model, Plan means balancing demand and supply. Plan also includes an organization's understanding of its available resources and constraints and how those available assets or limitations support or limit the

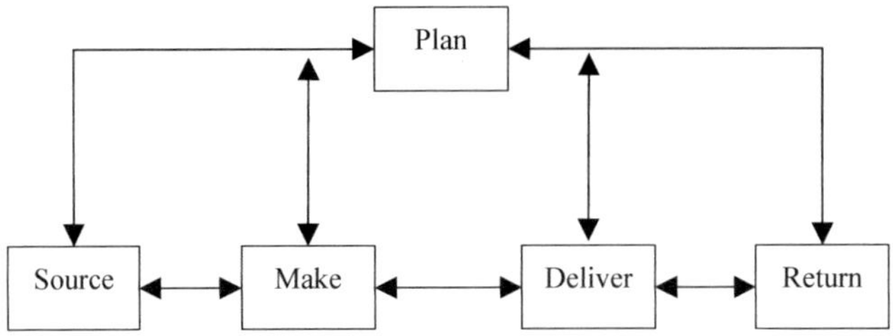

Figure 5.1 Loop view and the linear view of SCOR.

management of demand and supply. Planning is a cornerstone stage that will profoundly direct later stages, such as workforce capacity, inventory management, facility utilization, and customer services. It is the starting point of the SCOR Model and it is one of the stages in a loop of communication and feedback to determine new strategies, new services, and new products (Fig. 5.1).

The importance of planning activities is in its impact on the effectiveness of the supply chain and eventually on end users. How does an organization balance demand and supply? A synonym of Plan that supply chain professionals use is forecasting. In general, experts agree that accuracy of forecasting is the most important planning activity, and even the most important measure of a company's supply chain operation.

In a library setting, to balance demand and supply could be viewed as having too many collection volumes is no better than having too few. SCM is about accurate inventory control. It is not about adding to the collection just for the sake of adding and increasing inventory numbers. "Forecast," as a concept and practice, is contradictory to the established methods of academic and research library communities. Balancing demand and supply seems counterintuitive to collection services. Library acquisitions are about balancing budget and volumes. Collection services have been in a paradigm of building a sizable body of volumes by spending down the budget to zero dollars each year. The total amount spent on the collections is considered part of the investment libraries make. Forecast and collection development may appear contradictory; however, collection development, traced back to its origin, is a form of SCM forecasting. Historically, most libraries only acquired collections that support their university endeavors and seldom deviated from the research and teaching areas of their parent organizations. SCM "Forecast" always starts with user requirements in mind. Additionally, librarians select subjects of each discipline taught and researched on campus to fine-tune the collections. Often, it is arranged via the Approval Plans.

Libraries need to adjust, calibrate, and modernize their collection development techniques to make their forecast more accurate.

Libraries must use their funds more effectively and efficiently, as proliferating expectations of library support for research must be answered with flat budgets. The increasing budgetary pressure for emerging needs is a primary motivation to develop more accurate forecasting practices. Library stakeholders hold libraries accountable for making a significant impact on research and scholarship. To revamp their forecasting techniques is a must path toward improving collection values that showcase to their stakeholders their careful stewardship of collection resources. Librarians need to accurately forecast their user needs. The rate of accuracy must incorporate that of the collection usage. Usage is a longitude concept for libraries at the moment. Often, academic libraries collect extensively with an idea of building collections that reflect anticipated user needs in 10, 50, or 100 years; however, in SCM realm, many industries consider the concept of utility very differently from librarians. Having "assets" put aside for future uses is not an efficient way of stewarding financial resources because those funds end up not being allocated in ways that support current needs. Prioritization of library finances to balance present and future demands is an art and science that libraries need to master.

To accurately plan for user demands has been challenging for many industries. Forecasting is often done in ways of predicting and guessing based on the historical transaction data. In libraries, "prediction" and "guessing" are done based on the existing disciplines at their universities. Blanchard (2010) pointed out that 100% accuracy is unachievable, so the goal is to lower the error rate as much as possible. Error is common across some industries. For example, for the consumer industry, the forecast error typically ranges from 14% to 40%, with an average of 26%. The error rate means that retailers get it right 60 to 86 times out of 100. So where do libraries stand with their collection usage? Some studies on circulation showed the rate usually hovers around 33% in the next 10 years from the initial acquisitions time; however, during the first year, circulation is typically much lower and measures only between 10% and 17%. If benchmarked and compared with the retailers' forecasting errors, the libraries are only getting it right 10–17 times out of 100 in the first year and 33 times out of 100 in the next 10 years. In supply chain terms, the projection of collection needs is under the condition of "Over-forecasting," which results in creating more collection inventory than necessary. With large unused collection inventory, the costs of labor and storage will also rise in addition to the spent acquisition budget on those collection items.

As libraries are transforming themselves to be more customer-centered organizations by quickly responding to user needs, collections, as a core service, must also adapt and change. As collections take an enormous percentage of the overall budget, the value of collections should be rethought after all. Without the changes of collection strategies and techniques, commitment to customer-centeredness remains as an aspiration rather than a reality. Who can declare service excellence while stocks of items are marginally used? All the resources locked in the unused inventory could be redirected toward advancing emerging demands. The currently prevalent strategies hold precious resources captive to too many just-in-case scenarios. Library of today has to embrace just-in-time mode and go with a more demand-forecasting model. Chapter 6 will recommend specific methods that author believes will help libraries do a better job of collection planning and forecasting.

SOURCE—ACQUISITIONS AND LICENSING

"Source" also refers to procurement and purchasing. In the context of collection services, it probably can be easily translated into acquisitions and licensing activities; however, the SCM defines "Source" a little broader than what libraries may consider the concept and its included activities.

Based on APICS, "Source" describes how to manage inventory, the supplier network, vendor agreements, and supplier performance. It discusses how to handle vendor payments and when to receive, verify, and transfer products. Source involves all the infrastructure, processes, systems, and evaluation mechanisms to acquire necessary materials to meet planned demands. SCM supplier network and provider agreement are equivalent to the meanings of the vendor relationship and purchasing/licensing agreements in libraries, although, sourcing is more than what the Technical Services department manages in three perspectives described below.

Collection inventory, including backlogs and newly acquired and incoming collections, is part of the Technical Services purview; however, once the inventory is moved to the shelves, digital stores, or warehousing facility, inventory management responsibility transitions to the Public Services. The problem here is two-fold. First of all, the ownership of the inventory is not clearly defined in libraries; and secondly, the collection inventory is not a widely accepted concept in libraries. Lacking a precise ownership model often creates siloed and segmented planning between Technical Services and Public Services. Without an owner, infrastructure, processes, and systems become fragmented; human and financial resources are less effectively applied. What does collection inventory mean? When does it start and end? Does it begin when items show up at the loading dock and conclude with items being positioned on the shelves? In short, libraries do not consider their collections as inventory. In SCM, inventory covers parts and final products alike before products are shipped to users.

Supplier performance is even more novice of a notion to the library world. The library market is mature and highly relationship based. Therefore, the culture does not call for performance-based relationships. Kress and Wisner (2012) provided a way to measure vendor performances in product costs, their willingness to negotiate, reliability, responsiveness, and transparency; yet, there is little evidence from the literature that performance-based vendor relationship is widely adopted. SCM is about end-to-end integration with suppliers and customers. Suppliers are strategic partners to serve customers. Libraries should not measure performance solely on price and should not let vendors bear all the market risks. With that said, if vendors do not meet libraries or their patrons' expectations, vendors are not being great partners and put libraries and their patrons in inopportune situations as well. The true partnership involves a sophisticated selection process, through which library vendors and libraries establish long-term associations. This long-term cooperation builds a strategic relationship to align infrastructure, systems, and processes to maximize the benefits of the supplier, the library, and, ultimately, the end users. Sourcing will be elaborated on in more detail in Chapter 7.

MAKE—CATALOGING, DIGITIZATION, AND OTHER COLLECTION ENHANCEMENTS

Make not only includes the production of final products or services but also of such processes of packaging, staging, and releasing them. It also has to include production network, equipment, facility, and transportation. "Make" is perhaps a challenging SCM process for Libraries to comprehend in a traditional sense of library operations. Simply put, libraries usually do not "make" anything. Libraries buy or license materials from vendors and then make them accessible to users. Libraries neither package them nor transport them far elsewhere. Collections stay mostly local, and collections remain the same way vendors originally packaged them.

Make, as a SCM process, has to be substantially adapted to the library environment. "Make" can be seen as transformative activities and operations to either enhance or add value to collections. Factories "transform" individual parts into a functional product that meets user needs. Retail stores "transform" consumer goods through their selection, categorization, and consultation services. In a similar fashion, libraries go through a series of processes to "transform" books and other items through their selection, cataloging, digitization, preservation, curation, and reference services. Cataloging, for example, is an activity of "making," which adds valuable metadata either to provide a better search experience or better preservation. To generalize, any activities that enhance the discoverability or curation of collections can be regarded as SCM Make. Therefore, any facility, equipment, human resources, and information associated with those activities should be included in the Make processes. Make processes can even be a positive influence on such a trivial undertaking as transporting materials from the Technical Services area to the shelves.

Especially in the world of digital content, how libraries may acquire, curate, and provide access to them is entirely different than how they manage their physical collections. Digital materials require new "make" processes, equipment, resources, and expertise. Collection facility is a virtual space, and curation of collection requires new tools and procedures. Libraries must incorporate "Make" concept as they are expanding into Presses, Open Access publishing, and Open Source development.

"Make" may provide libraries with new perspectives on the values that they bring to their patrons. It offers an opportunity to begin a new path to examine library-established workflows, procedures, and processes. The Michael Porter's Value Chain Model, introduced in the same chapter, frames the following question for libraries: what value do library activities bring to users? For example, many libraries embraced shelf-ready collections option because they realized that cataloging common collections might not be the best use of expertise and resources. Shelf-ready collections free up catalogers to work on other priorities. Because all collection enhancement work fits into the Make processes, libraries ought to question many status quos and look for alternative ways to get their processes streamlined. "Make," if done well, will bring library operation in alignment with the current and future technological context and user trends. Chapter 8 will explain the Make process in more detail.

DELIVER—STACK SERVICES AND COLLECTION DELIVERY

Deliver encapsulates all the activities that transport the finished product to consumers. It typically includes customer ordering, invoicing, warehousing and life cycle management, and transportation (could be both international or domestic).

Given the unique situations found in libraries, "Deliver" is the processes that the libraries have to modify considerably to fit into their operation. A majority of libraries provide document delivery service to certain demographics of their patrons, if not for all. SCM Delivery processes can translate into collection delivery service including management of orders, facility, collection life cycle, and transportation. Even considering the self-retrieval services in the open stacks, libraries may see shelves as a destination/delivery point, Delivery to Stacks. Delivery to Stacks can significantly benefit from the clearly defined and established collection life cycle and facility management practices of SCM "Delivery" processes. Whether it is Delivery to Stacks or Document Delivery, libraries may apply SCM principles to reexamine the processes and procedures to ensure service quality.

Collection logistics, such as facility and storage management, will be essential for service quality. Delivery can start as early as acquisitions (Source), go through cataloging (Make), and eventually arrive to the shelves or be sent to users. Delivery to Stacks will have to offer findability and browsability by allocating resources for frequent and time-consuming stack arrangement. Digital materials do not deal with the same issues as physical shelves, but they demand resources to establish a new kind of infrastructure, systems, tools, and processes to ensure discoverability and delivery. As more libraries are moving toward a just-in-time service model, Delivery will greatly impact users satisfaction. Libraries will need to design a delivery process that can originate from the supplier to carrier services and span to their internal workflows. The speed of delivery depends on the efficiency of the overall supply chain. In Chapter 9, "Deliver" will be developed in more detail.

RETURN—COLLECTION RETURN

Return is part of the customer service support. In any case of returning products, a company must prepare to handle defects, unsatisfactory products. "Return" involves the rules, policies, facility, packaging, inventory, and transportation. In libraries, it often means that users return borrowed material back to libraries for future use. Occasionally, libraries may need to return acquired materials in poor condition back to vendors, but those are rare occasions compared with those of library user returns. Perhaps the concept of "Return" is simple and straightforward, even with consortia borrowing arrangements and interlibrary loan services; however, timely managed returns with well-designed workflows and planned facilities and logistics contribute to better service quality. Chapter 10 will focus on SCM "Return" process in libraries.

As Kress and Wisner (2012) pointed out in their article, "A supply chain model can assist in delivering value for library users at a reduced cost through the integration of services. Thus, SCM can aid in meeting the personalized requirements of library users, while using lean systems to make efficient use of scarce economic resources for libraries as content providers." Later chapters in this section will dive deeper into the SCOR Model processes and metrics and their applications in libraries.

REFERENCES

APICS. (2015). *SCOR framework*. APICS Supply Chain Council. Retrieved from: http://www.apics.org/sites/apics-supply-chain-council/frameworks/scor.

Ball, D., & Wright, S. (2000). Procuring electronic information: New business models in the context of the supply chain. *Library Consortium Management: An International Journal, 2*(7), 145–159.

Blanchard, D. (2010). *Supply chain management best practices* (2nd ed.). Hoboken, NJ: John Wiley & Sons. Wiley best practices.

Chopra, S., & Meindl, P. (2001). *Supply Chain Management Strategy, Planning, and Operation*. Upper Saddle River, NJ: Prentice-Hall.

Cornish, G. (1996). Librarians: Peacemakers in the new electronic world. *Library Management, 17*(4), 30–33.

Huan, S., Sheoran, S., & Wang, G. (2004). A review and analysis of supply chain operations reference (SCOR) model. *Supply Chain Management: An International Journal, 9*(1), 23–29.

Hudson, S., & SCRC (2014). *The SCOR model for supply chain strategic decisions. NC State Poole College of Management Supply Chain Resources Cooperative*. Retrieved from: https://scm.ncsu.edu/scm-articles/article/the-scor-model-for-supply-chain-strategic-decisions.

Kress, N., & Wisner, J. (2012). A supply chain model for library quality and service improvement. *Journal of Operations and Supply Chain Management, 5*(2), 40–53.

Meng-xing, H., Chun-xiao, X., & Yong, Z. (2010). Supply chain management model for digital libraries. *The Electronic Library, 28*(1), 29–37.

Mourits, M., & Evers, J. J. (1995). Distribution network design. *International Journal of Physical Distribution & Logistics Management, 25*, 43–57.

Perez, H. D. (2013). *Supply chain strategies: Which one hits the mark? (CSCMP and supply chain. media, LLC), Supply Chain Quarterly*. Retrieved from: http://www.supplychainquarterly.com/topics/Strategy/20130306-supply-chain-strategies-which-one-hits-the-mark/#fn7.

Porter, M. (1980). *Competitive strategy: Techniques for analyzing industries and competitors*. New York: Free Press.

Porter, M. (1985). *Competitive advantage: Creating and sustaining superior performance*. New York: Free Press; London: Collier Macmillan.

Wang, Zheng (John) (2013). Co-curation: New strategies, roles, services, and opportunities for libraries in the post-web era and the digital media context. *Libri, 63*(2). (print).

CHAPTER 6

Plan—Collection Planning

As Chapter 5 indicated, Plan involves balancing demands and supplies and matching user requirements with available resources. Although Plan is just one of the processes of the supply chain operations reference (SCOR) model, it really is an overarching process. Plan activities should cover and inform other SCOR processes: Source, Make, Deliver, and Return, and it should touch on the integration aspects between suppliers and customers. An organization must start with Plan to build an effective and efficient operation.

This balancing of demands and supplies is crucial simply because there is not an entity that has all the resources to sustain every single one of its needs. Plan always matches all types of resources, such as personnel, facility capacity, cash, and equipment, to all available products and services. An organization needs to determine the life cycle of their products or services and then allocate resources for achieving their short- and long-term goals. One of the key activities of Plan is to forecast needs.

COLLECTION PLANNING AND SCM FORECASTING

To accurately forecast, a firm first needs to understand the market position and context of each product. Likewise, libraries need to recognize the market context and position of their services. Although forecast can apply to all services, such as reference, instruction, and research data management, this chapter only covers collection services.

Libraries historically do not plan services from the perspective of the supply chain management. Budget, staffing, and other resources often gravitate toward maintaining enduring commitments without desired flexibility to manage emerging demands. As purchasing power is shrinking as a result of annual cost inflation and growing deficits, it becomes even more dire to sustain current services. Any expansion to meet new user requirements frequently has to be requested from parenting organizations, which often results in a multiyear effort without any guarantee of sustainable and long-term funding increase. Forecast offers an opportunity for libraries to support both existing and emerging needs within their current resources situation.

Supply Chain Management for Collection Services of Academic Libraries
ISBN 978-0-08-102031-9
http://dx.doi.org/10.1016/B978-0-08-102031-9.00006-2

Collection planning and SCM forecasting may appear in conflict, but the goal of collection planning is similar to that of the SCM forecasting. The difference is that SCM planning requires periodic assessments of changes and thus agility in redesign of the chain processes, whereas collection planning is often carried out through acquisitions and collection development processes that have not seen any fundamental changes in decades. The set of methods and strategies remains intact even though the society had long shifted from the environment in which they were originally conceived and formulated.

Traditionally, a big percentage of collection budget goes toward a collection method, called Approval Plans. Those plans are automated purchases based on established subject profiles. It takes a form of collaboration between vendors and libraries, where library sets up the profiles while suppliers fulfill those plans and ship materials to libraries. Those approval materials typically come in large quantities and often require much support from Technical Services; however, as a paradox, collection planning often does not include capacity planning for Technical Services. To address the bottleneck of processing Approval Plans, more libraries have embraced and paid for the "shelf-ready" collections option, in which materials ship with cataloging records and spine labels; however, collection planning often does not consider the need for quality assurance of those records.

Most critically, the profiles in vendor systems are not adjusted as frequently as necessary and their systems vary greatly in flexibility and ability to fine-tune them as libraries desire to develop deep collections over time. One of a major inherited shortcomings of the Approval Plans was the large body of unused collection items, which begs for the question of the return on values. Is it the best way to spend the budget, labor resources, and other investments into Approval Plans? Approval Plans prioritize possibility of use over the real-life demands from user communities. The current low usage of collections generates an opportunity for cost savings and redirection in libraries. The investments, such as personnel, facilities, and utilities allocated toward long-term curation and preservation, could instead be channeled toward management of emerging needs. If those trapped resources spent on infrequently used materials are freed up, libraries could gain the required flexibility and agility to meet the emerging needs, such as startup research tracks, and support of newly established academic disciplines.

The opportunity cost is a real issue for all collection plans that prioritize collection-building over collection-use. Granted, firm orders and

Interlibrary Loan service address gaps created by blanket purchase plans; however, they are small in expenditure compared with those large programs. From the perspective of SCM, the established collection planning processes do not adequately address demands using available resources. Resources are allocated for existing accounts that include serials, monographs, and e-resources, and thus, there is little left to accommodate anything else. The planning process crashes libraries' ability to respond to patrons' new requests. The current methods resulted in a massive overage of collection inventory. The author advocates for a complete overhaul of collection development as a strategy. Not only was the entire current system established for analog collections but also, more importantly, its effectiveness has been markedly diminishing in service to users within digital context.

The new environment that libraries operate in requires responsiveness and agility. SCM proposes a set of new processes that establish a platform for libraries' collaboration with patrons and vendors and a path to provide high-quality collection services. Before introducing SCM Plan, it is necessary to help readers understand the SCOR model in depth, particularly its attributes and level-one metrics and processes.

SCOR LEVEL-ONE PROCESSES AND ATTRIBUTES

Plan, Source, Make, Deliver, and Return are five level-one processes of the SCOR model. To measure the SCM efficiency, SCOR recommends use of performance metrics based on five attributes: reliability, responsiveness, agility, cost, and asset management. The first three, reliability, responsiveness, and agility, are customer-focused attributes; the other two, cost and asset management, are internal attributes (SCC, 2010).

Libraries first need to establish a supply chain strategic direction based on the SCOR attributes. The strategic direction guides libraries to implement an appropriate instance of supply chain and create prioritization for their collection process improvement. A library may choose to lower costs and build up assets, a common practice of renewing or canceling collection packages.

Taking Assets in libraries, as an example, it is represented in an extensive inventory of collections. Whether collection size is reflective of service excellence is a myth in today's context, although size has been one of the long-standing metrics that guided libraries' "collecting" behaviors. The size was a significant indicator of success when publication, discovery, and delivery were geographically bounded and constrained by the modes of transportation across a network of libraries.

In SCM, Asset is an entirely different notion. SCOR model (SCC, 2010) describes Asset as "the ability to utilize assets efficiently. Asset Management strategies in a supply chain include inventory reduction and insourcing versus outsourcing." As content discovery and delivery are increasingly ubiquitous and instant, users value relevance over quantity. Can libraries provide services that are reliable, responsive, and agile for all patrons and all disciplines? Size just cannot serve all users, regardless of how deep a collection might be. Because resources are limited, libraries cannot amass an extensive collection for every discipline on campus. Flagship collections for selected disciplines do not translate into service excellence across the board.

The Supply Chain is a proven strategy for improving customer service outcomes, so it is important for libraries to prioritize their collection services by introducing SCOR customer-focused attributes. Cost and Asset Management should be orientated toward achieving Reliability, Responsiveness, and Agility. All three aforementioned SCOR attributes are values that libraries are striving to blend into their cultural DNAs today. SCM offers libraries opportunities to implement them at operational level across all functions and services.

Reliability addresses the predictability of the outcome of a process (SCC, 2010). Typical metrics for Reliability include on-time, right quantity, and right quality. Responsiveness describes the speed at which tasks are performed. Typical metrics include on-time and cycle-time. Agility describes the ability to respond to external influences and the ability to change. Agility is measured by flexibility and adaptability.

"On-time" is one of the key metrics of Reliability and Responsiveness. If along the supply chain, tasks are designed and managed promptly, services will be delivered on time. In other words, the process performance aggregates to the service timeliness.

What does "on-time" mean for collection processes and services? Libraries could gauge the time that lapses between the moment that a user identifies the need for an item to the moment he or she receives it, whether the item is physical or online. For example, users may request acquisition of a data set that requires legal counsels, electronic resources librarians, subject liaisons, and procurement staff and IT staff to secure the use right and licenses, process the payment, and mount and activate the data set with special access rights. Can current collection workflow accommodate those types of unexpected needs in a timely manner? How long does this work take? From the author's personal experiences, those requests often take months or, in some extreme cases, even years to solve. On-time means quick turnaround to meet patrons' needs. When they plan collections, libraries must consider all the following three scenarios: long-standing needs, emerging needs, and ad hoc demands.

Right quantity and right quality are quite new notions for libraries. Libraries traditionally consider quality as the depth, completeness, and specialty of collections. From collectors' point of view, a high-quality collection translates into high volume of material. SCM forecast balances resources with needs of users. Quantity and quality of services are positively evaluated only if relevant research materials can be discovered, accessed, or delivered in a timely matter. With the advancement of cross-disciplinary research, an in-depth collection can still be porous, not adequately serving a particular research or scholarly need. It is time for libraries to adopt SCM concept, forecasting the right quality and right quantity for every discipline on campus.

Currently, print collection usage is low, which means libraries are overforecasting user demands. Overforecasting wastes resources in the initial acquisitions and ongoing maintenance of infrequently accessed materials. The goal of the planning or forecasting is not to get to 100% accuracy, rather in lowering the error rate. Only after libraries truly understand the demands of users, they can do better planning work. The Amazon Use Cases shared in Chapter 3 noted that Amazon differentiates frequently sold goods from other merchandise. Libraries should start planning for collection segmentation, patron's segmentation, and market supply analysis.

COLLECTION SEGMENTATION

Libraries will need to address all three scenarios: daily collection requests, emerging collection needs, and ad hoc needs. The author suggests the following segmentation of three dimensions for collection planning (Fig. 6.1):

1. Necessity: Must-Have and Good-to-Have
2. Usage: Frequently Used, Occasionally Used, and Never Used
3. Uniqueness: Special Collections and Common Collections

The Must-Have category contains primary resources that a department needs to have access to for it to support its scholarly or research activities. Subject liaisons should contact their departments to understand those needs rather than self-determining the needs. The collection could include reference works that departments frequently use, journals that departments read

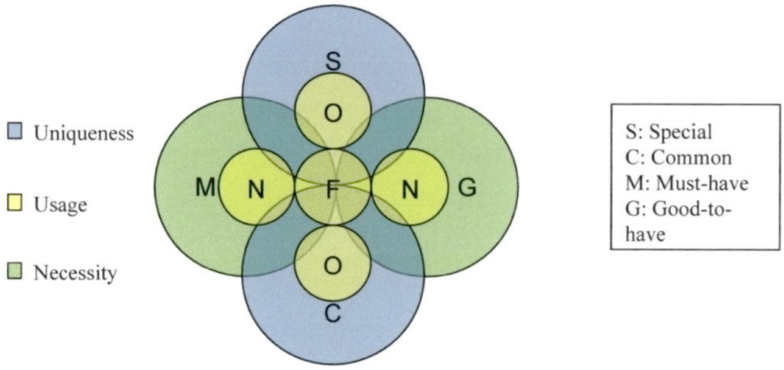

Figure 6.1 Collection segmentation.

to keep up with the new developments in their areas, or data sets or other resources that departments require to conduct their day-to-day work.

Good-to-Have collections are anything that patrons see as optional. Libraries should consider just-in-time access to or delivery of those collections. It is essential to collaborate with vendors to plan Good-to-Have collections. Libraries may inform vendors about their projected needs based on usage data. Libraries and vendors should work together to figure out pricing and a workflow model to deliver content for the sake of end users. As the above critical information is exchanged, both vendors and libraries will improve their ability to forecast quantity and quality of new publications.

Another dimension of the Segmentation is usage. In addition to working with patrons to determine "Must-Have" collections, libraries may use circulation data to confirm or challenge either their or their patrons' assumptions about what is necessary. The usage statistics help libraries to not only better augment what "Must-Have" and "Good-to-Have" collections are but also allocate resources to unanticipated collections. Instead of having a fixed budget for each discipline, libraries may develop a new budget model that offers much flexibility to support anticipated and unanticipated needs. Usage-based segmentation and planning could be very politically charged because the new budget model will change the control and ownership of budgets. In Section Three, Future Proof Library Supply Chain, the writer will discuss soft issues associated with SCM implementation in libraries.

Uniqueness is the third segmentation of the Collection Planning. Special collections are assets, as their value is likely to increase over time. Those type of collections should be universally considered rare in their availability and value to research, scholarship, and education. The three segmentations of necessity, usage, and uniqueness should work together to achieve planning efficiency and ultimately deliver excellent collection services to users. Excellence is an aspiration and a possibility that could be achieved through the partnership of like-minded institutions and their partners, including vendors. Those segmentations do not conflict with the desire to have in-depth and complete collections. A "delayed" buying technique can be implemented as one of the techniques because price drops tremendously overtime.

PRIMARY MODE OF ACCESS

Plan should also consider Primary Modes of Collection Access, as the SCM strives to improve services for users at the end of the chain. Should libraries plan for users to retrieve materials directly from the stacks or deliver items to users, or both?

It is a core question that libraries must determine the direction for their collection services today and the future. Perhaps the best way to answer this question is to demystify a couple of assumptions that librarians have about patrons, or even patrons might have about themselves. For example, a certain percentage of faculty often claim the absolute importance of availability of the stacks to their scholarship; however, they also state how long they have not been to the stacks as they frequently request that collections be delivered to their offices. The "productivity" of stack browsing is diminishing, as more collections only exist in electronic forms. The stacks will not offer a complete browsing experience for faculty as more acquisitions are added to the entire body electronically. As faculty petition how much they prefer the serendipitous discovery that stacks offer, the truth is that virtual browsing offers much better serendipitous discovery than physical shelves. Acknowledging that only some faculty will still elect to browse physical collections, librarians should understand the changes in faculty research behaviors and advocate for better means and services to support teaching and research faculty needs. Libraries need to educate faculty about the fact that no single library can offer collections that will meet all scholarship needs, so the stack browsing is no longer as sufficient as it used to be. The fact is more faculty praise the Document Delivery Service and more faculty use online databases and electronic journals, delivery to their office, personal computers, and mobile phones, all of which are becoming a norm.

The majority of students will prefer benefits of online access as well. Recently, the author personally received a student's email request for digitizing all collections in the library to improve productivity of research and learning of the campus student body. He mentioned his experience of stack browsing was counterintuitive and compared it with his experience of browsing HathiTrust digitized books. More students are accustomed to the online access as a primary mode of access. A percentage of current students will become scholars and researchers in the not-so-distant future. Because of that, libraries must prepare their collections primarily for delivery. For this reason, the author will use delivery as the primary mode of access; however, the author will also suggest workflows and metrics that track performances of possible retrievals in the stacks.

COLLECTION PLANNING STRATEGIES AND METRICS

The author recommends "Make-to-User" and "Make-to-Stack" as two primary strategies to plan collection services. Make-to-User means that libraries work with vendors to streamline their supply chain processes for quick collection delivery directly to users. This strategy is good for "Good-to-Have" and "Occasionally Used" collections. When users place a request or an order, libraries and vendors work together to deliver the item quickly, irrespective of whether the libraries own the item.

Make-to-Stack means the supply chain processes view shelves as the end point of services. Users find items on shelves on their own or retrieve them with the assistance of library staff, depending on whether they are special or general collections. This strategy is intended for Must-Have and "Special Collections." Interlibrary Loan and Document Delivery Services are considered derivative services of the collections in the stacks. Interlibrary Loan and Document Delivery follow their own processes and performance measures, which are not part of the two strategies, particularly not of Make-to-User.

The author recommends distinctive processes for the above two Strategies, but their performances will be measured based on the SCOR attributes: Reliability, Responsiveness, Agility, Cost, and Asset Management. Based on SCOR definition (SCC, 2010), metrics for the level-one processes, Plan, Source, Make, Deliver, and Return, are strategic metrics, called Key Performance Indicators (KPIs). KPIs are diagnostics for the overall health of a supply chain. The author recommends the following KPIs for the two strategies (Tables 6.1 and 6.2):

Make-to-User KPIs

Table 6.1 Make-to-user KPIs

Collection services supply chain attribute	KPI
Reliability	1. Perfect collection delivery 2. Perfect collection download
Responsiveness	1. Delivery cycle time 2. Download time
Agility	1. Collection upside flexibility 2. Collection adaptability
Cost	1. Collection supply chain costs 2. Collection and support costs
Assets	1. Collection utility 2. Utility on collection supply chain fix assets 3. Utility on working capital

Make-to-Stack KPIs

Table 6.2 Make-to-stack KPIs

Collection services supply chain attribute	KPI
Reliability	1. Search match with collection inventory 2. Stack item findability
Responsiveness	1. Stack cycle time
Agility	1. Collection upside flexibility 2. Collection adaptability
Cost	1. Collection supply chain costs 2. Collection and support costs
Assets	1. Collection utility 2. Utility on collection supply chain fix assets 3. Utility on working capital

Reliability and Responsiveness KPIs are self-explanatory. The two Strategies share the Agility, Cost, and Asset Management KPIs. The section of Collection Planning and Forecasting introduced the concepts of Reliability, Responsiveness, and Assets Management. This section on KPIs gives introduction of the notions of Agility and Cost, as their definitions guide users learning about KPIs.

Agility is an attribute used to respond to sudden, enormous, and unforeseen changes. SCOR (SCC, 2010) lists examples of external influences, including nonforecasted increases or decreases in demand; supplier or partners going out of business; natural disasters; acts of (cyber) terrorism, availability of financial tools (the economy); or labor issues. Agility in supply chain comprises flexibility and adaptability. Flexibility measures time taken to respond to a demand change, either increase (upside) or decrease (downside). Adaptability measures the maximum quantity of production variations in a given time.

In library setting, Agility is the ability of libraries to respond to unexpected and significant changes in collection needs. Those changes could be a request to establish a new collection for an emerging academic area or subject, acquire data sets to conduct new research, or even downsize a collection of a no longer existing department. Flexibility is the reaction time to those unplanned changes. Adaptability is the capacity of the libraries to process content requests or orders at peak and low times. Those peak times are often at the beginning or end of a semester; low times are often other periods of a semester, such as spring break, fall break, or summer break. At peak times, libraries need to plan for sourcing, enhancing, and delivery of collections, for example, Course Reserves, interlibrary loan (ILL), and Firm Orders. At low times, libraries can typically handle routine volume of requests or orders for sure, but the planning piece during the low times should include preparation of workforce and resources for other duties, services, or priorities.

Collection Supply Chain Cost measures total supply chain management cost, which includes the fixed costs and those costs of operations associated with the collection processes. SCOR is a hierarchical model, which nests three tiers of processes and metrics. Collection Supply Chain Cost is an aggregation of costs from underlying processes. Libraries should consider all costs associated with collection services, including HR, systems, equipment, facility, utility, collection funds, etc.

Again, Assets in the context of a supply chain are in the management of their efficiency. Metrics describe the ability to use them efficiently. Because the Assets attribute favors efficiency, its KPIs are one type of indicators that demonstrate return on investment. Libraries often see Assets as size and completeness of their collections. The Assets KPIs challenges this traditional practice, instead proposing the efficiency of utilization of resources.

Libraries also need to define the Cycle time of Assets. In SCM, Cycle time of the Assets is formulated in days, while libraries should consider setting their Cycle time on a yearly basis. Libraries should set a goal of a certain percentage of collection utilization every year. Some libraries may take into account usage over 5- or even 10-year periods. If libraries are courageous, can they perhaps set a goal of 100% usage every year? Libraries may also consider collection usage over the cost of its curation, such as labor and total collection budget. Only through those lenses, libraries can understand the return on investment of all their content. Without those types of KPIs, it would be challenging for libraries to change their collecting behaviors and thus difficult to free up resources to manage emerging needs.

With the understanding of some basic concepts of this chapter, Chapters 8–11 will drill down to the hierarchy of the SCOR model and link lower level of processes and their metrics with the strategic KPIs of collection services.

REFERENCE

SCC. (2010). *Supply chain operations reference (SCOR) model overview – version 10.0.* Cypress, TX: Supply Chain Council, Inc.

CHAPTER 7

Source—Acquisitions and Licensing

CURRENT ACQUISITIONS AND LICENSING IN LIBRARIES

In libraries, Source, based on the supply chain operations reference (SCOR) model, is the acquisitions and licensing processes. Those are often part of the Technical Services (TS). TS is an enduring function of any traditional library. Although the name of the department has been evolving and additional functions are being added to reflect the emerging formats of collections, the processes have not been coming along at the rollout speed of new formats and new standards. The processes remain stable and include ordering, shipping, check-in, record management, account payables, etc. Sourcing for electronic content almost mirrored the exact processes used for physical collections, except for shipping. Those steps have been part of a larger process for a long time and remained largely successful; however, the sustainability of such a model is questionable as evident by backlogs. One may argue that backlogs are cataloging issues, which in SCM term is the Make process. If libraries take a systematic view of collection management, Acquisitions and Licensing contribute to cataloging backlogs. Often, depending on the Acquisition team staffing level and the volume of incoming collections, backlogs are generated already from the very first step of ordering thus creating a domino effect that propagates through the rest of the supply chain.

Notably, the service culture of TS has not seen much change since its inception. TS has been a somewhat self-contained area—what goes into it will come out eventually, but often at its own pace, on its own schedule, and based on its own expectations and prioritization of work. How collections are processed, in what order they are treated, and how long it may take for materials to be released for patron use were often unknown to the rest of the libraries' departments. Because of the technical specificities and system sophistication, not many others outside TS, such as subject liaisons and library administrators, understand the details of TS work. Other departments typically trust TS for its work to be completed; at the same time, such a situation created lacking transparency and operations of TS with highly departmentalized decision-making of the "fate" of its own department and that of collections.

Supply Chain Management for Collection Services of Academic Libraries
ISBN 978-0-08-102031-9
http://dx.doi.org/10.1016/B978-0-08-102031-9.00007-4

The author is not suggesting that TS abuses their autonomy, just proposing that such departmental culture impacts the service bottom line and hurts TS in general. In the author's view, the speed of "Source" affects all five attributes of the SC performance, in the order of responsiveness, reliability, agility, cost, and asset management.

The level of autonomy hurts TS because of its opacity to outsiders. Because collection decisions anecdotally are made without the involvement of TS, TS is often unprepared either at the skill level or at the staffing level to process those collections. Collection managers tend to buy collections without any consultation with TS or preparation for processing of the new purchases. Whatever they buy, they assume that TS has or will come up with a way to find the needed expertise and capacity to handle their purchases and those newly acquired collections will be ready for their constituencies. Because of the lack of understanding from top management and administrative advocacy, TS has been given retired positions to immediate pressing needs, such as research, technology, and digital. In some cases, libraries hire metadata librarians, whose expertise is a particularly critical part of the know-how that TS needs to expand to; however, they are often added to other functions or newly formed entities, such as Scholarly Communications, Digital Libraries, Metadata units, etc. With time, lack of transparency created a vicious cycle: collections are growing exponentially, and formats are getting more complicated, and yet staffing and resources in TS are shrinking and unable to successfully manage the workload increase and types of work change. The ultimate loss is reflected not on TS, but on libraries and their users.

To change this phenomenon, library leaders should start integrating TS processes with the rest of library collection services. Supply Chain concepts and models can provide new perspectives to the current TS leaders in their awareness of linking service excellence with TS process improvement. Supply Chain models offer an opportunity to train TS leaders to review the basis of TS operations and create meaningful workflows and metrics that would serve end users.

LIBRARY SOURCE PROCESSES

Source is one of the five level-one processes of the SCOR model, describing the scope and high level of configuration of a supply chain. It includes all processes that interact with vendors and deal with receiving materials from suppliers. SCOR (SCC, 2010) defines it as follows:

> *The Source describes the ordering (or scheduling) and receipt of goods and services. The Source process includes issuing purchase orders, scheduling deliveries, receiving, shipment validation and storage, and accepting supplier invoices.*

In a library setting Source can be mapped to collection acquisitions and licensing process. The acquisitions and licensing include tasks of ordering and receipt of library collections. Some commonly known assignments are ordering, claiming, receiving, check-in, invoice, and payment for analog collections; negotiating, testing, record-loading, activating, invoice, and payment for electronic collections.

SCOR is a hierarchical model. Each level-one process is further expanded down to several levels. Based on its manual, SCOR suggests standardized processes from the Level One down to the Level Three, at which processes are still widely applicable across multiple industries. SCOR allows an organization to extend the Model to lower levels to include proprietary processes. A process at each level is a unique activity performed to meet predefined outcomes. Higher level processes can decompose to lower level processes; lower level processes can aggregate to higher level processes.

Level Two processes differentiate the strategies of the Level One processes. Both the Level Two processes themselves and their positioning in the supply chain determine the supply chain strategy (SCC, 2010). The Source processes should be guided by three Segmentations of Collections: Necessity, Usage, and Specialty, and Make-to-User and Make to-Stack strategies explained in Chapter 6. This chapter establishes a set of principles as guidelines that will aid in developing library-specific source processes and metrics in the later part of the Chapter:

1. Serve both existing and emerging collection needs
2. Balance the above needs with available resources
3. Apply collection segmentation to order the right quantity and quality
4. Prioritize electronic over physical, but offer print-on-demand
5. Prefer "Make-to-User" over "Make-to-Stack"

The set of strategies ensures proper planning of sourcing that takes into account all user collection needs in an aggregated way. The planning process based on SCOR should involve steps to:

Identify, prioritize, and aggregate collection requirements
Identify, assess, and aggregate collection resources
Balance collection resources with collection requirements
Establish sourcing plans

The better an organization understands its user needs, the better its Source processes are. The better an organization understands its organizational capacity and resources, the better its Source plans are. Source planning will determine who libraries should work with, what to buy from each vendor, how many volumes to buy from each vendor, and how much it costs to buy from each vendor. The author recommends involving vendors in Source Planning, considering the Extended Supply Chain. The goal is to help vendors understand user needs and resources availability. Libraries should view vendors as strategic partners, and vice versa. The Source strategic relationship means that vendors are not interchangeable parts in the supply of knowledge to users, rather it helps define a long-term partnership that informs vendors to position and shape their understanding toward user needs. By doing so, it becomes a win–win situation to eliminate waste for both vendors and libraries, and foremost for users. As any projection of requirements may incur inaccuracy, Source Planning processes will need to be built into regular intervals to understand the changes in demand.

Once the Planning is complete, Source Execution process starts. Execution is a process triggered by Source plans or actual demands that change the state of collection volume. How the Source Plans are executed contributes primarily to the Stack Cycle Time, Delivery Cycle Time, or Download Time. Execution processes are the processes of acquisitions and licensing tasks. The value of examining the set of processes through the lens of the Supply Chain is to recommend an architecture of processes that can be clearly measured for their benefits to end users. Interestingly, Cycle Time or Download Time is not a metric that libraries currently implement to gauge service excellence, particularly not within TS departments. TS views their service excellence in precision and completeness of bibliographic records, rather than the timeliness of completion of each task and the efficiency in connecting those tasks. Although libraries have been exploring ways to improve acquisitions or licensing processes, the practices and endeavors may greatly benefit from the guidelines outlined in the Supply Chain model.

In Chapter 6, the author introduced two SCOR Level-Two Processes, Make-to-Stack and Make-to-User which differentiate the Level-One Processes. Source execution processes at the Level Three should also align with the two strategies, as the lower processes aggregate to higher level ones and contribute to overall performances. At the Level-Two, each execution process has three different possible capabilities of representing and responding to customer orders (SCC, 2010). Each category also affects Plan and Return processes. "Make-to-Stack" is adapted from a typical Level-Two process, Make-to-Stock. Make-to-User is adapted from the commonly known process Make-to-Order. The third one, Engineer-to-Order is not a focus of this book, but could be a strategy adopted by other services of the libraries, such as research data support.

Make-to-Stack model guides Plan, Source, and Make processes to delivery of collections to library physical and virtual stacks. Libraries get collections to the stacks, and then patrons locate those collections on their own. Applying the Collection Segmentation is crucial in this capacity. The stocked collections have to be must-have and frequently used, or they should be specialty collections. The Segmentation is the only way to balance existing and emerging needs with a library's perpetually fixed and flat budgets. The stacked collection in this way can free up the budget for other needs and save budget on maintenance of storage of noncirculating collections. The Make-to-Stack model builds on the foundation of inventory control and turnovers. Although libraries do not sell collections to their patrons, inventory control and turnover can be regarded as the utility of collections and collection facilities. Source Execution processes must guarantee repetitive uses of collections for both short term and long term.

Make-to-User means a collection item is not ordered until a patron makes a request. The advantages are that the collection is configurable based on user requests, and the majority of the budget is reserved until the actual order is placed. Make-to-User model has been implemented as the Demand-Driven Acquisitions (DDA) or Patron-Driven Acquisitions (PDA). DDA or PDA have been leveraged by libraries to manage library budget in a way that is different from traditional methods. There have been several variations or iterations since PDA entered into library markets. It has taken such forms as short-term loans and triggered purchases after a certain number of page views or lease of the entire collection and retention of the used titles at the end of the lease period. Make-to-User is a perfect strategy for good-to-have collections, and occasionally used collections, as a way to supplement Make-to-Stack. Those good-to-have and occasionally used collections are currently prevalent in library holdings, not only taking up valuable storage space but also locking funds that could be allocated for priorities and areas of emerging demands that libraries do not have budget to handle. The author here does not ask libraries to blindly adopt PDA models, as some of the models might not be in the best interest of end users or create efficiencies in Supply Chain tasks. Libraries need to implement Make-to-User strategy and processes holistically. As more electronic collections are delivered to users through PDAs and DDAs, users can access those collections immediately; however, it does not mean that the workflow constructed for the delivery of content to users is optimal at this time. Make-to-User may be a slower method than Make-to-Stack; yet, as long as patrons understand the rationale of this method for meeting their broader needs, the author believes that users would be amenable to having reasonable turnaround. The author recommends the following Source processes for libraries and understands customization is desired because of the distinctiveness of individual library operation:

Source for Make-to-Stack (Source-for-Stack) processes
- Order collection
- Schedule delivery
- Receive collection
- Verify, claim, and check-in collection
- Transfer collections to catalog or storage
- Authorize vendor payment

Source Make-to-User processes (Source-for-User)
- Load vendor publication data
- Activate publication data
- Verify orders
- Order items
- Schedule deliveries (for analog collections)
- Receive items (for analog collections)
- Transfer items to catalog or user
- Authorize supplier payment

Although Source-for-Stack processes might be similar across libraries, Source-for-User tasks require further customization among libraries. Libraries have options to receive and verify the orders first and then decide to process or deny the orders, as the above processes suggest. Or, libraries may decide to allow users to place an order and then ship the order directly to users:

- Load vendor publication data
- Activate publication data
- Order items
- Schedule deliveries (for analog collections)
- Transfer items to user
- Authorize supplier payment

Because the environment of each library varies, the author will leave the decision on policy implementation and further customization details to each individual institution; however, no matter whether libraries or users receive ordered content, there is a chance that vendors might send wrong materials or defective products. In those cases, libraries will need to design a process to return those materials. The author will also leave creation of those processes to each library to configure and will discuss the importance of creating processes that concern performance.

SOURCE PERFORMANCE METRICS

The above-suggested Source processes will link up to higher level performance metrics and eventually connect to KPIs. For example, the "Make-to-Stack" processes contribute to the Responsiveness KPI, Stack Cycle Time in the following structure:

Level One—Stack Cycle Time

Level Two—Source Cycle Time

Level Three

—Authorize supplier payment cycle time
—Transfer collection to catalog or storage cycle time
—Verify, claim, and check-in collection cycle time
—Receive collection cycle time
—Schedule delivery cycle time
—Order collection cycle time

The time needed for the above-listed six tasks links to the overall Source Cycle Time and then to the Stack Cycle Time. The lower level metrics provide diagnosis for higher level indicators. For example, if the KPI, Stack Cycle Time indicates a significant delay, libraries can drill down to lower levels to identify the reason. Libraries might find out that the handover from the acquisitions to cataloging or storage was slow or that the verification and receipt of collection was slow. Either way, the metrics help libraries to find the inefficiencies and hence improve the workflow.

Source workflow also contributes to Costs KPIs. Study of Sourcing Costs will also give libraries insights into the utility of their resources and thus may optimize the Supply Chain processes to optimize the use of their limited resources. Taking Collection Supply Chain Costs as an example:

Level One—Collection Supply Chain Costs

Level Two—Costs to Source

Level Three

—Cost to authorize supplier payment
—Cost to transfer collection to catalog or storage
—Cost to verify, claim, and check-in collection
—Cost to receive collection
—Cost to schedule delivery
—Cost to order collection

Although the author does not give examples of how Cost to Source links to Collection and Support Costs, it is obvious that Source workflow including process design, staffing level, equipment, and system all directly impact collection services at the end. Normally, libraries do not see cost factors in the context of Supply Chain, but rather break down cost into separate discrete areas, such as integrated library system cost, Online Computer Library Center (OCLC) cost, and staff cost. Therefore, large investments are often made in isolation. Supply Chain Costs Metrics start to link individual cost components together and thus offer a holistic view on library investments for the end results.

REFERENCE

SCC. (2010). *Supply chain operations reference (SCOR) model overview – version 10.0*. Cypress, TX: Supply Chain Council, Inc.

CHAPTER 8

Make—Collection Enhancements

MAKE IN LIBRARY SETTING

Make in Supply Chain Management (SCM) describes manufacturing and production activities. Usually, the Make step includes production activities, packaging, staging, and releasing. It also includes managing the production network, equipment and facilities, and transportation (SCC, 2010).

This chapter integrates Make into library "manufacturing and production activities." The Make step occurs after the Plan and Source steps. At this point in time, collections have arrived at the libraries, so now libraries will need to make the collection accessible, discoverable, and long-term usable. In general, one may consider Make steps for libraries as any collection enhancement work, including cataloging, metadata, digitization, conservation, and preservation work. One library may be part of a consortium or a network of libraries to collaborate on the Make steps. University systems with many campuses or regional coalitions of libraries are examples. Whether independent libraries or a network of libraries, they will all need equipment and facilities to carry out their Make steps. Transportation of the Make steps is trivial for most libraries, as their collections are mainly "made" locally and serve local constituencies. Library networks that own shared facilities should look into transportation workflow management, since their Make processes are spread among several geographical locations; for others, transportation merely means routing between main library and branches on campus, from Make units, such as technical services, preservation, digitization to and from the branches.

Chapters 6 and 7 covered two major Make strategies: Make-to-User and Make-to-Stack. Make-to-Stack is geared toward must-have and specialty collections; while, Make-to-User processes are tailored to good-to-have collections. Libraries will have to treat the Make processes of the two strategies differently. Make-to-Stack will be a traditional way of cataloging, holding work, and conservation and preservation after the collections arrive at the libraries. More libraries have ordered shelf-ready collections for general collections. The cataloging may not be needed, but loading of records is still an essential step of the workflow. With special collections, the process might

Supply Chain Management for Collection Services of Academic Libraries
ISBN 978-0-08-102031-9
http://dx.doi.org/10.1016/B978-0-08-102031-9.00008-6

be different from the workflow of general collections, in that they will still need the whole process of cataloging, holding, and loading work. Libraries should study the process carefully for potential improvement over time.

Make-to-User is a relatively new method that libraries have been embracing more these days. The strategy manifests itself in patron-driven acquisitions and demand-driven acquisitions types of operations. In Make-to-User, cataloging records are preloaded to the library systems. The process is unique. The cataloging records are preset by suppliers, so the quality of those records determines the discovery of those resources. Librarians often prefer the workflow to customize pre-records either for search or long-term preservation purposes. This set of workflows has room for improvement as well.

Make steps significantly contribute to the Responsiveness of the Collection Supply Chain. The time takes to complete the Make steps, "Make Cycle Time," aggregates to Responsiveness key performance indicator (KPI), Fulfillment Cycle Time. Since the Make steps are aggregated from lower-level Make processes, how quickly an organization can cycle through those lower-level processes also contributes to the Make Cycle Time metric. In SCOR, Make Level-Three activities and their measures are shown as examples below:

Level One—Order Fulfillment Cycle Time

Level Two—Make Cycle Time

Level Three

— Finalize Production Engineering Cycle Time
— Issue Material Cycle Time
— Produce and Test Cycle Time
— Release Finished Product to Deliver Cycle Time
— Schedule Production Activities Cycle Time
— Stage Finished Product Cycle Time
— Package Cycle Time

LIBRARY MAKE PROCESSES

Applying the above in library setting, libraries will need to tweak their processes significantly. The most important element to adapt is "time" as a metric. For organizations like libraries, "time" is not often a measure used to judge quality of services. The backlog of collections is a prevalent problem among many libraries. Backlogs, under SCM context, mean poor responsiveness, which is an indicator of low-quality services.

Since the backlogs are often sealed from the patrons, in other words, they can only discover materials that are searchable; patrons are not aware of potential impact of the backlogs on their research. Since backlogs are investments but are not discoverable; libraries must solve this issue. In this sense, time is an important measure. Recognizing the uniqueness of each library, the author only suggests the following model of Make processes for a typical library:

Processes for Cataloging/Metadata

Arrange collection for cataloging
Perform copy cataloging
Perform original cataloging
Assure Record Quality
Load records
Stage enhanced collection
Release enhanced collection to storage

Processes for Conservation/Preservation

Remove items from storage
Transport to the lab
Repair items
Stage repaired items
Release items
Transport items to storage

Processes for Digitization

Remove items from storage
Transport to the lab
Digitize items
Edit digitized items
Update records
Link records with digitized surrogates
Transport items to storage

The above processes leave room for libraries to customize as needed based on their individual environment. The important thing to do is first to answer questions why those Make processes are needed, and why each of them is unique and necessary. Ask why it is extremely critical for the library industry, given many of collection procedures have been in place for decades since integrated library system (ILS) was invented, and some have existed for print media even for a longer period of time. After review, some of the steps can be retired, so libraries won't spend time on gauging responsiveness of obsolete steps. The following demonstrates how metrics are nested, using Digitization as an example:

Level One—Stack Cycle Time or Download Time

Level Two—Make Cycle Time

Level Three

— Transport Items to Storage Cycle Time
— Link Records With Digitized Surrogates Cycle Time
— Record Updates Cycle Time
— Edit Digitized Items Cycle Time
— Digitize Items Cycle Time
— Transportation to Lab Cycle Time
— Remove Items From Storage

LIBRARY MAKE METRICS

Make steps also contribute to the Agility of the Collection Supply Chain. The agility of supply chain is measured in the ability of an organization to respond to sudden changes. Libraries may think of those sudden changes in the forms of rush orders, unpredicted last minute needs. Under those circumstances, can libraries be agile enough to respond to patrons' needs promptly?

One of the KPIs in this category is the Upside Supply Chain Flexibility. It is a discrete measurement defined as the amount of time it takes a supply chain to respond to an unplanned 20% increase in demand without service or cost penalty (KPI Library). Translating this definition in libraries setting, do the libraries have a supply chain in place that is flexible enough to manage 20% surge over an average volume of work? The service or cost penalty could be seen as the delay in collection discoverability. In SCOR, Upside Make Flexibility, a level-two measure, contributes to the Upside Supply Chain Flexibility:

Level One—Upside Supply Chain Flexibility

Level Two—Upside Make Flexibility

It is important to note that Preservation, Conservation, and Technical Services are often not the units that see growth. In fact, they are those areas of the organization that suffer from shrinking Full Time Employees (FTEs) that are more often than not allocated to other emerging areas. The above practice impacts libraries' flexibility to handle any unexpected surges as discussed above. It creates a vicious cycle—as libraries buy more collections, less Make capacity to process them on time is available. Matching Source

Plans with Make capacities is critical from the supply chain point of view. The author is not suggesting that libraries should increase those Make units since not everything libraries are currently buying is valuable to patrons from the supply chain perspective. The first step is for libraries to fundamentally change their mindset on collection building.

Agility, second KPI, is Supply Chain Adaptability. In supply chain, Adaptability is a way to ensure that a business may adjust Make output so it can sustain a substantial increase of orders or large reduction in need at 30 days mark. This, on the surface, seems like it has nothing to do with libraries; however, adaptability should be a value of library Make steps. From time to time, libraries are facing increases or reduction of collections, for example, addition of a large gift collection. Again, libraries have a fixed size of the technical services, staff, and preservation units that they have to work with to adjust to large volume changes. Even more so, libraries' Make processes are also fixed and are not particularly responsive to shifts in user demands. While patrons' needs for collections subside, what often happens is that libraries have no reductions in sourcing and supplies of collections. So an extensive inventory of collections that was created is now sitting on physical or digital shelves. Should libraries respond to user desires and adjust their Source and Make volumes? If reduction were to happen, how can a library plan FTE for other types of work? For example, perhaps print books are in decline, but data is on the rise. Can library shift Make units to process data? Those are the set of questions that libraries need to answer today.

It is valuable to have the assurance that Make workforce capacity can handle the increase or decrease without impact on the current priorities of the libraries. Twenty percent flexibility and 30-day adaptability are arbitrary numbers provided for benchmarking purposes, according to SCOR (SCC, 2010). Different industries have options to choose meaningful benchmarking standards. Can libraries handle 5% increase or sustain substantial increase or decrease in collection volumes of 5-day work? Flexibility and adaptability may seem like a luxury for libraries at this time, but the author believes they can be realized by adapting SCM concepts.

Level One—SupplyChain Adaptability

Level Two—Make Adaptability

What is the cost of Make then? Have libraries asked such a question to understand their investment in those Make steps? Make Costs contribute to the Collection Supply Chain Costs. The Make Costs can be a significant part of the overall supply chain costs, when a library calculates the labor, equipment, IT, and other necessities that support the Make steps.

Level One—Collection supply chain costs

Level Two—Make Costs

Level One—Collection and Support Costs

Level Two—Make Costs

The last one is the Make Fixed Asset Value, which also is an essential element that contributes to the Utility on Collection Supply Chain Fix Assets. Utility on Collection Supply Chain Fix Assets measures the return an organization receives on its invested capital in supply chain fixed assets. Since libraries are not-for-profit organizations, revenue is not the type of measure that libraries need. Chapter 7 proposed use of collection utilization as a way to represent returns on investment. Since all steps along the supply chain contribute to the returns on investment, libraries need to understand the fixed assets of their Make steps as one piece adding to the overall returns. When a library looks at its labor costs from collection volumes and usage, a library will have a whole new understanding of the values of technical, preservation, digitization, and conservation functions of the libraries.

Level One— Utility on Collection Supply Chain Fix Assets

Level Two—Utility on Make Fix Assets

Level One—Utility on Working Capital

Level Two—Utility on Make Working Capital

REFERENCES

KPI Library. (n.d.) Upside supply chain flexibility. Retrieved from: http://kpilibrary.com/kpis/upside-supply-chain-flexibility.

SCC. (2010). *Supply chain operations reference (SCOR) model overview – version 10.0*. Cypress, TX: Supply Chain Council, Inc.

CHAPTER 9

Deliver—Stack Services and Delivery

DELIVER FOR MAKE-TO-USER

The Deliver processes describe activities associated with creation, maintenance, and fulfillment of customer orders. It includes the receipt, validation, and creation of customer orders; scheduling order delivery; pick, pack, and shipment; and invoicing the customer (SCC, 2010). In the business setting, delivery includes order management, warehousing, and transportation. It also includes receiving orders from customers and invoicing them once the product has been received. This step involves management of finished inventories, assets, transportation, product life cycles, and importing and exporting requirements (SCC, 2010).

This definition of Deliver needs to be adapted to the library setting, as libraries do not invoice but rather keep lending records and do not usually deal with overseas readers but rather serve local users. However, libraries handle processes that include requests/orders, pick, pack, schedule, and ship items; even more so, libraries may borrow SCM school of thought to manage collection inventory and collection life cycle.

The definition of Deliver for libraries should be applied to the Make-to-User strategy introduced in Chapters 6 and 7. Make-to-User is a strategy where libraries collaborate with vendors to streamline their supply chain processes for quick collection delivery to users directly. This strategy works well for "Good-to-Have" and "Occasionally Used" collections. Instead of buying such items and stacking them on shelves, libraries preload catalogs of various publishers and only trigger a purchase when a user requests an item. For a user, Deliver starts with search, find, and request; for libraries, Deliver starts with receiving, processing, tracking, and sending the "ordered" content. The requested content can be delivered online or via traditional shipping methods.

Also, anecdotally, library patrons browse shelves far less frequently than they used to even a decade ago. Circulation of stacked collections is profoundly low across many libraries, compared with online discovery and

Supply Chain Management for Collection Services of Academic Libraries
ISBN 978-0-08-102031-9
http://dx.doi.org/10.1016/B978-0-08-102031-9.00009-8

access. The above observation indicates user emerging preferences, which should guide libraries to direct their service improvements and resources allocation to Make-to-User strategies.

SUGGESTIONS FOR INTERLIBRARY LOAN AND DOCUMENT DELIVERY

Because the "Make-to-Stack" processes present shelves as the endpoint of services, this chapter will not cover this strategy. The Make-to-Stack strategy is designed for "Must-Have" and "Special Collections." The above collections should be acquired and ready to use anytime. In the author's opinion, Interlibrary Loan (ILL) and Document Delivery services are considered derivative services of the collections in the stacks.

For the reason that ILL and Document Delivery follow their own processes and performance measures, this chapter will not cover them in detail but mention some possible improvements that ILL and Document Delivery may borrow from SCM thinking. Processes of ILL and Document Delivery are the closest to business delivery services, in which libraries need to deal with order management, warehousing (collection facilities), and transportation. It is critical that libraries manage the book inventories, knowing what assets there are and what is lendable, understanding life cycles of the collections, and tracking receipt of the collections. In some cases, ILL and Document Delivery deal with delivery across country borders. So the ILL and Document Delivery steps can be closely measured; however, this is not a part of services that libraries measure with great preciseness. The author recommends that libraries measure ILL and Document Delivery processes. The author experienced many cases of the "lost" items once they were sent to users. This situation often happened because items were delivered to wrong places. Many users either forgot updating their campus addresses or gave a shared work space with others. First of all, the "lost" items are detrimental to collections and resources; Second of all, they are negatively impact ILL statistics—labor and resources went into the processes without any outcome. Think about the work steps or processes that need to be put in place to address those issues. Given ILL and Document Delivery is becoming more dependent on by faculty members, libraries should investigate and invest more effort in understanding the under-the-hood operation of those services. Measuring processes in a detailed fashion will help determine when something goes wrong and thus provide opportunities to improve delivery processes. Another problem is the asset management—what can be borrowed physically and what should never ship out the library door. There are many steps and parties involved in ILL and Document Delivery. If a mistake is made along the way as part of any step or by any party involved in the process, an item can be lost and never returned. The author suggests libraries design processes for checking the worth of the content and determine delivery, whether via original copies or digital surrogates. With that said, the rest of the chapter will focus on processes and metrics for the "Make-to-User" strategy.

DELIVER PROCESSES AND METRICS FOR MAKE-TO-USER

Online collection delivery is instant, the type of measurement of "online delivery" will be entirely different than that for the physical delivery. For this reason, the author recommends two sets of processes for the Make-to-User, but libraries should customize the processes based on their own environments:

Processes for Electronic Items

Level One—Deliver

Level Two—Make-to-User

Level Three

Process requested items

Validate items

Purchase or lease items

Deliver items

Processes for physical items

Level One—Deliver

Level Two—Make-to-User

Level Three

Process requested items

Validate and order items

Consolidate items

Purchase items

Receive items from vendors

Consolidate deliveries

Pack deliveries

Route deliveries (on campus)

Select carriers and rate shipment (out of campus)

Ship deliveries (out of campus)

Receive and verify deliveries by patrons

Libraries may also build in more levels into the above processes if libraries need to define their delivery processes further. Take "Purchase or Lease" item, for example:

Level One—Deliver

Level Two—Make-to-User

Level Three

Process requested items

Level Four

Compare pricing information from vendors

Involve subject liaisons to determine scholarly value of items

Determine purchase or leasing

Because those lower level processes can be very specific to libraries, the author will not list them all, but will note that those levels roll up to the higher level process performance. Supply Chain metrics will allow libraries to pinpoint concrete steps for diagnosing issues.

As earlier chapters discussed, supply chains are measured by their five characteristics: reliability, responsiveness, agility, cost, and assets. Deliver processes contribute to supply chain Key Performance Indicators.

The author recommends two metrics for reliability, "Perfect Collection Delivery" and "Perfect Collection Download," discussed in Chapter 6. Those top metrics are decomposed to Deliver Level-Two Metrics:

Metrics for Physical Items

Level One—Perfect Collection Delivery

Level Two

Delivery performance to service standard

Perfect delivery documentation

For physical items, Level-Two metrics can be diagnosed by Level-Three metric:

Level One—Perfect Collection Delivery

Level Two

Delivery performance to service standard

Perfect delivery documentation

Level Three

Perfect customer receiving

Delivery location accuracy

Metrics for electronic items

Level One—Perfect Collection Download

Level Two

Near–real-time delivery

These hierarchical metrics are self-explanatory. If an imperfect collection delivery occurred, libraries would drill down to find out where the hiccup happened. As long as libraries logged the receiving dates and addresses, libraries can check the delivery performances from the lower level metrics. Libraries should extend to more levels of metrics as necessary.

Author recommends Level Two and Level Three metrics for responsiveness below:

Responsiveness Metrics for Physical Items

Level One—Delivery Cycle Time

Level Two—Deliver Cycle Time

Level Three

Receive and verify deliveries by patron's time
Ship deliveries (out of campus) time
Select carriers and rate of shipment (out of campus) time
Route deliveries (on campus) time
Pack deliveries time
Consolidate deliveries time
Receive items from vendor's time
Purchase items time
Consolidate items time
Validate and order items time
Process requested items time

Responsiveness metrics for electronic items

Level One—Download Time

Level Two—Download Cycle Time

Level Three

Delivery items time
Purchase or lease items time
Validate items time
Process requested items time

By linking those processes and measures, libraries will be able to diagnose a particular process and determine any performance issues.

Deliver also contributes to Agility, Cost, and Assets Management. In the earlier chapter, the author introduced flexibility and adaptability. Flexibility is a concept dealing with unplanned collection increase without incurring more resources, such as hiring temporary workers. Adaptability is dealing with increase or decrease in demand with the current resource and staffing level. Deliver Flexibility and Adaptability, of course, contribute to the overall Supply Chain Flexibility and Adaptability. Understanding of

Deliver Flexibility and Adaptability can help a library run Deliver unit workforce better.

Library should calculate personnel, systems, software, carrier service fees, and overhead expenditures associated with Deliver processes to know the overall Supply Chain costs. Understanding of Deliver cost can help library guide Deliver resources.

For asset management, libraries should know their "return on supply chain fixed assets." In the delivery sense, the fixed assets are those that libraries invest in their Deliver processes. This number could be used to aggregate to a higher level to understand the total return. Libraries will get an average of collection usage over their total investment in all five top-level processes.

REFERENCE

SCC. (2010). *Supply chain operations reference (SCOR) model overview – version 10.0.* Cypress, TX: Supply Chain Council, Inc.

CHAPTER 10

Return—Collection Return and Get Ready for the Next User

RETURN IN LIBRARY SETTING

Return in supply chain management (SCM) describes processes that companies must be prepared to have in place when handling the return of containers, packaging, or defective product. The return involves the management of business rules, return inventory, assets, transportation, and regulatory requirements (SCC, 2010). Although libraries do not have to deal with many of the complications as for-profit companies do, library Return involves the management of circulation rules, return stocking and reshelving, assets, transportation, and regulatory requirements, such as fines.

Although the above are the major use cases of Return, it is likely that from time to time libraries have to handle the return of wrong orders or wrong items identified either by patrons or by staff, reversing the supply chain direction all the way back to vendors. Libraries have very specific steps, processes, and workflows to handle return activities, whether they are returns of collection items to the libraries or returns of wrong orders back to vendors.

Return seems not as critical to libraries as Plan, Source, Make, and Deliver, the other four processes of the supply chain. Generally speaking, patrons are not returning something defective. They are borrowing and returning collections that are useful to them. The urgency is not as high as that of proprietary companies who handle returns as part of their business plans, and libraries' reputation is not at stake at the same level as of those companies. After all, collections are returned back after they have served the patrons, unlike Return in other industries where Return occurs when the goods are not meeting customer needs. Scan of library science literature shows very little research on return management: an evidence of lack of interest and lack of attention to the topic in the profession. Since it has existed as a necessary step, librarians might need to stop and think about how Return may impact overall service quality. Often libraries look at such measures as total circulation and satisfaction surveys to gauge overall patron satisfaction; however, those measures do not uncover inefficiencies or issues

Supply Chain Management for Collection Services of Academic Libraries
ISBN 978-0-08-102031-9
http://dx.doi.org/10.1016/B978-0-08-102031-9.00010-4

at the workflow level that actually contribute to overall satisfaction. Neither do they highlight issues and problems that contribute to user lack of satisfaction. Libraries just do not tie Return, or any of the Supply Chain processes, with their service quality, in general.

This does not mean that libraries should neglect their return processes. Currently, the other four processes are definitely more important than Return, in author's opinion; however, Return processes also link and contribute to the quality of user services. This is true maybe not so much for patrons who return items after checking them out, but for returning patrons or prospective patrons who need the same items. In the future, Return processes can become extremely important, as publishers are figuring out better methods of applying copyright law to library lending of electronic publications. In section Return Measures of this monograph, the book will introduce some innovative ideas that may redefine Return processes for libraries.

RETURN MAKE-TO-STACK OR MAKE-TO-USER

As discussed in earlier chapters, Return is a SCOR Level-One process, which can be further drilled down to lower-level processes. SCOR recommends three levels of processes, and individual institutions may further define the levels that fit their organization structure. The book also discussed the second level of SCOR processes and strategies that each entity may adopt to support their supply chain. The author recommends that libraries approach their collection with "Make-to-Stack" and "Make-to-User" strategies. "Make-to-Stock" converted to library industry is "Make-to-Stack"; while, "Make-to-Order" for libraries means "use triggered collection development strategies, which in libraries are Patron Driven Acquisitions (PDA) and Demand Driven Acquisitions (DDA)." In Chapter 8, the book also discussed the Make processes for libraries and defined them as metadata enrichment and collection enhancement steps that increase the discoverability or preservation quality for immediate and long-term use of library collections. For PDA or DDA, libraries often source the discovery metadata from vendors; however, the preservation functions are still owned by libraries, which means that libraries still need "Make" steps to manage preservation metadata and preservation work. The following chapter will provide typical level-three return processes and metrics for libraries. Libraries should use it as an example, since their own steps may vary based on institutional needs.

The Return steps for both the "Make-to-Stack" and "Make-to-User" can be modeled the same, regardless whether the collections reside in digital stores or on physical shelves. For physical items, whether they are ordered from Source Plans or ordered based on user requests ad hoc from the "Make-to-User" Strategy, the steps for return are the same. The only caveat is that when an item is recalled, specific steps should be put in place to ensure that next users can get the item quickly. Otherwise, Source steps should be set to order more copies of the same title by going through the "Make-to-User" steps.

Currently, for digital content, the Return steps, if there are any, are straightforward to users. For many collections which do not have digital rights management (DRM), users have simultaneous access and no return is needed. In cases where digital content is being assigned with DRM, return is often automatic as loan periods expire on the digital platform where those collections are hosted. Currently, such automatic processes are more happening with electronic books rather than electronic journals and articles. Some of the ebooks are part of databases which are licensed for concurrent access; part of the ebooks are modeled after their physical counterparts, where a copy is unavailable if it is checked out. With open access content gaining popularity, electronic collections will become more complex. The author does not feel that there will be an ideal model that will work for publishers, libraries, and end users soon. As libraries are becoming smarter shoppers that look at the value that their acquisitions budget brings to their patrons, the author thinks one of the options for libraries could be a shift to more "Make-to-User" type of SCM for electronic collections. The shift from subscription to individual purchase might reset the electronic content models so they mirror models of the print, where a purchase of a copy will only allow a single user at a time to have access. No matter how it may change, Return steps are becoming less of an issue for library supply chain.

Level One—Return

Level Two—Make-to-User or Make-to-Stack

— Reclaim Returned collection
— Reinventory Returned Collection
— Check-in Recall Status
— Sort Returned Collection
— Restock Returned Collection, or
— Reroute for Delivery of Recalled Item

From time to time, libraries will need to deal with defective items or even wrong shipments from vendors. What libraries acquire is beyond just collection items themselves these days. Acquisitions often include bibliographic records and other accompanying artifacts. Any of the above could be considered "defective." Walker and Kulczak (2007) did a study on shelf-ready books. They found out that 27.05% of records exhibited errors affecting catalog access, and records for 38.59% of titles needed modification. Moreover, 32.11% of series headings required further attention to comply with local authority procedures. From the supply chain point of view, 27%–38% is a huge defect range. Errors in records impact the ultimate user discovery of the collections. More so, they also impact the stakeholders' expectation concerning the return on investment: if libraries spent money on collections and made them shelf-ready, shouldn't they be discovered and used, and shouldn't they save library time and effort freeing staff from cataloging and labeling them for the shelves? If a quarter of the collection could be undiscoverable due to poor record quality, is it the best use of libraries' resources? Often, the approach for handling shelf-ready errors is to have staff to "double process" them for better discovery, which completely defeats the purpose of shelf-readiness. This is only one example. Shelf-ready record quality has improved since its early days, but the example illustrates a bigger issue, which is the shipment of discovery metadata by publishers. Because publishers and aggregators are very proprietary on metadata, when libraries are aggregating them into a single platform, such as their catalog or discovery layer system, the discoverability of those collections may not be searched fairly and equally by users. Although libraries pay good money for all their collections, not all of them can be easily found by users. Even publishers realize that because of the variation in search and metadata, when their metadata becomes aggregated in a central index, the elevation of their collections discoverability does not necessarily translate into more users discovering their content. This is a significant problem for publishers, libraries, and users. Back in the day, those issues might have been handled by libraries returning the physical items. In electronic world, it is very difficult to return electronic products. Not all book sellers are allowing their customers to return ebooks; if allowed at all, typically, it has to be done in a very short window of time after purchase. In the world of electronic content, Return process is less ideal. So the entire Supply Chain must be reimaged to accommodate challenges the new medium presents. The new Chain should be established to improve qualities of discovery services associated with those collections. There are many innovations that could be implemented between publishers and libraries to revamp those processes.

Thus, it is important to understand the workflow, the resources, and time effort put toward managing those unfortunate situations, as their management can contribute to the overall responsiveness, flexibility, and cost of the Supply Chain. Particularly, defective items tend to happen during "Make-to-User" collection services. At least initially, libraries should consider learning the metrics that associate with the workflow, know how those relate to those of collection management, and use the information as a way to determine how much effort the libraries should invest or how soon the libraries need to invest into optimizing Return processes. This assessment can vary among different size of libraries. Overall, the processes are the same as those in other industries. A typical example could be:

Return of defective collections by libraries

Source return defective

— Identify defective collection condition
— Dispose defective collection
— Request Return
— Schedule shipment
— Return Defective collection
— Check refunds or confirm replacement

RETURN MEASURES

Return, as the last stage of a Supply Chain, does not impact Reliability as directly as other four processes; however, in a case of collection being delivered to patrons through Make-to-User, Return metrics should help libraries see whether the delivered items were matched with user requests and whether they were defect-free. Libraries should aim for 100% accuracy in those two metrics, so no return would be necessary; however, if errors were to occur, it is also important to use at least two metrics that could point libraries to the right path toward improving their processes. Those metrics are designed specifically for physical collections, but libraries may use them for their electronic collections as well, even though the chance for delivery of wrong or defective electronic items is fairly low.

Reliability

For physical items:

Level One—Perfect Collection Delivery

Level Two—Perfect Condition

Level Three

— Percentage of Matched Requests
— Percentage of Defective-Free

How return is handled has direct impact on service Responsiveness and Agility. The author recommends that libraries track time of each Return task:

Responsiveness

Level One—Return Cycle Time

Level Two

— Reroute for Delivery of Recalled Items Time
— Restock Returned Collections Time
— Sort Returned Collection Time
— Check-in Recall Status Time
— Reinventory ed Collection Time
— Reclaim Returned Collection Time

Agility is defined the ability to manage huge increase or decrease of user demands. Agility metrics offer libraries a way to balance their resources and staffing levels with some possible demand scenarios over time. Author does not believe that libraries will experience surges of Return, either in defective returns or in normal collection returns; however, it is an assurance for libraries to know confidently that their staffing and resources can handle X% increase, sustain substantial increase in return volumes in X number of days, or reassign Return staff from substantial decrease in Return volumes in X number of days without allocating more resources or staff. The author recommends that libraries track defective returns and collection returns separately, since the meanings of the two types of "Return" are different.

Level One—Upside Supply Chain Flexibility
Level Two—Upside Return Flexibility (Source Defective)
Level One—Upside Supply Chain Flexibility
Level Two—Upside Return Flexibility (Deliver Defective)
Level One—Upside Supply Chain Flexibility
Level Two—Upside Return Flexibility (Collection Return)
Level One—Supply Chain Adaptability
Level Two—Return Adaptability (Source Defective)
Level One—Supply Chain Adaptability
Level Two—Return Adaptability (Deliver Defective)
Level One—Supply Chain Adaptability
Level Two—Return Adaptability (Collection Return)

Cost to Return should also be tracked as Cost to Defective Return and Cost to Collection Return. Cost to Defective Return can associate with Source or Deliver activities. Libraries may even break down the metrics into lower-level metrics as needed. The value of those metrics is to understand the costs of "rework," and help libraries improve the processes to prevent later errors. Cost to Collection Return provides the insight into the resources for managing normal returns via both physical or electronic channels. Sometimes, Return may happen due to planning errors and order management. There is value in creating a metric to track performance of those processes.

Level One—Collection Supply Chain Costs

Level Two—Cost to Return (Defective)

Level Three

— Cost to Defective Return (Source)

— Cost to Defective Return (Deliver)

— Cost to Collection Return

Level One—Collection Supply Chain Costs

Level Two—Cost to Plan

Level Three

— Cost to Plan (Return)

The author recommends similar approach for Asset Management with libraries tracking resources and staffing and additional investments for other processes. The author sees the value in understanding Return investment in the context of collection utility.

Level One—Utility on Collection Supply Chain Fix Assets

Level Two—Utility on Return Fix Assets

Level One—Utility on Working Capital

Level Two—Utility on Return Working Capital

REFERENCES

SCC. (2010). *Supply chain operations reference (SCOR) model overview – version 10.0*. Cypress, TX: Supply Chain Council, Inc.

Walker, M., & Kulczak, D. (2007). Shelf-ready books using PromptCat and YBP: Issues to consider (an analysis of errors at the University of Arkansas). *Library Collections, Acquisitions, and Technical Services*, *31*(2), 61–84. http://dx.doi.org/10.1016/j.lcats.2007.07.016.

PART 3

Future Proof Library Supply Chain

CHAPTER 11

What Is Essential and What Is Peripheral for Libraries

SOME TASKS CAN BE OUTSOURCED

Libraries have been outsourcing many tasks, processes, and operations to third parties. Some of the outsourced work includes cataloging, digitization, and programming. Let's taking cataloging, for example. Reasons for outsourcing cataloging can range from lack in language expertise to lack of capacity to increasing size of backlogs. Davis (2014) summarized several efficiency and cost-control strategies that libraries have been implementing. One of such efforts is staff reductions resulting from sharing and outsourcing of cataloging, reference, and digitization services.

However, Davis did not see the above approaches translate into equal or better services to patrons. In his opinion staff reductions clearly have not created the needed budgetary flexibility to deal with rising cost concerns of the industry. Notably, as the price of publications keeps rising and currency continues to inflate, libraries have been fighting against a perpetual constant that exists outside of library control. So it is quite clear that many challenges just cannot be solved under their current operating philosophy unless universities centrally commit to sustaining library operations in the foreseeable future. Then why do libraries keep outsourcing supportive functions and downsizing their staff? The answer cannot be only due to economic reasons, but it is very likely guided by a bias on the mission criticality of those supporting departments: since they are just not as core as other functions of the libraries, they could be downsized!

Under the supply chain principles, all functions are essential components of the chain and equally contribute to the quality of end products or services. By examining the five top processes of a supply chain, plan, source, make, deliver, and return, plan of collections is achieved through collaboration between subject liaisons and acquisition staff; technical services manage a lion's share of source and make, and public services primarily handle deliver and return processes. Any of the processes of the five phases will lead to the quality of collection services at the end of the chain. Libraries should not consider outsourcing before properly understanding the capacity of

Supply Chain Management for Collection Services of Academic Libraries
ISBN 978-0-08-102031-9
http://dx.doi.org/10.1016/B978-0-08-102031-9.00011-6

their supply chains. The capacity of plan, source, make, deliver, and return should accurately match the flow of the content.

People sometimes mistake tasks of a function for the function itself. For example, cataloging is a means to describe collections, while collection description is a function. In addition to MARC21 traditionally used by catalogers, many other metadata standards can be utilized to describe collections. Collection description is part of the make processes. For libraries, make is the set of processes that add value to the acquired or licensed content from vendors. Speaking of the libraries' mission in discovery and preservation, Make is a crucial phase that libraries should not take lightly. Libraries have firmly established important standards, practices, procedures, and workflows, which are all part of make processes and are considered significant contributions to the knowledge industry.

This chapter does not argue that libraries should not outsource make processes at any point. Libraries should consider shifting their limited make resources to high value work. As value of certain methods diminishes, they can become good options for outsourcing. For example, libraries buy catalog records (quality control or defects is a separate issue) for shelf-ready collections, which is an excellent strategy for freeing up staff resources for other types of work. Outsourcing can also be a way to make up for the bottleneck resource capacity, for example, when local digitization demands exceed the capacity of existing personnel performing the task. Making blind decisions on outsourcing work or focusing strictly on support units as good choices for outsourcing is a mistake.

LIBRARY CORE EXPERTISE

As the section "Collection Enrichment and Preservation" of this chapter provides fundamentals of supply chain and its applications in collection services, this chapter hopes to develop a new understanding of essential functions of collection services from the perspective of the supply chain. Core is a business concept, a process through which to understand what to keep as unique expertise and what to acquire from contracts to or partnerships with vendors. When it comes to collections, libraries often prioritize along the following two lines:

- Ownership over processes and capacity
- Size over utility

The direct results of above guidelines are affecting support units: unprocessed backlogs, at-risk unpreserved materials, and loose collection inventory control. The ultimate impact is projected on user productivity in teaching, research, and learning. If collections are services, the following should be part of the services: the content itself, search interface to locate the content, curation of the content, metadata describing the content, preservation of the content, and facility for the content. If collections are services, libraries should connect people with the right content they need. Libraries should determine whether collection supply chain is the be-all and end-all model or if it only involves tasks that prioritize the content itself.

What is library unique collection expertise? To answer this question, libraries should ask themselves what differentiates them from other information providers. How does their unique expertise contribute toward user benefits? Libraries must introspectively examine each collection task closely and honestly. Once libraries understand their core, they may then choose what work to outsource and what tasks should stay. The chapter will share a few thoughts about what is core and non-core for collection services. In the author's opinion, libraries, as intermediaries between knowledge and patrons, play an essential role in content curation, enrichment, and preservation.

USER RELATIONSHIP MANAGEMENT

Curation, first and foremost, is a means to manage user relationship. It does not imply simply buying collections but should be represented by buying collections informed by the knowledge of user needs and providing sophisticated care for those collections. To curate successfully, libraries must understand collection requirements developed through careful review of relationships; yet in reality, collection acquisitions are often executed without taking into account such relationships. Without the in-depth understanding of user demands and using them as guidelines, libraries have to rely on broad teaching, learning, and research goals to conduct their collection development activities. In Esson, Stevenson, Gildea, and Roberts (2012), the authors recommend library services of the future should engage with customers to determine wants and needs. They point out that "the future-focused academic library must be distinguished by the scope and quality of its service programs. To be successful the design and development of those services have to be shaped and informed by the customers."

One of the author's former positions was in a web services department. It was there when he realized the criticality of User Experience (UX) design to the virtual library space for connecting users with the right content. UX techniques are widely utilized to understand user preferences for web applications and can also be leveraged for obtaining collection services needs. Bell (2014) envisioned that UX design moves from the periphery to the core of academic library operations. Some libraries have adopted customer-centered design frameworks into their environments.

UX or the service quality is the ultimate goal of the supply chain. Supply chain, regardless of how readers interpret the meaning of the term, is a user-centered philosophy. Although supply chain management advocates for process excellence, it has an end state in mind, which is expressed in service results. UX requires that libraries allocate more resources and time toward interaction and communication with their patrons, through which they would acquire deeper understanding of their collection requirements. Currently, the UX of collections has been widely perceived as "collection size, depth, and breadth," a practice that is seen as a high mark of quality service. A rich and complete collection is a high quality collection, but it does not translate into a high quality service, especially because the terms "rich" and "complete" are often relative rather than absolute expressions.

Large collections should never be the end goal for collection services from the supply chain point of view, which might have been a model for serving users in the past when the entire world operated on an analog paradigm; however, the advancement of technologies has completely changed user expectations. Continuation of collection building without consideration of other aspects of service quality can no longer provide adequate service to users. Collection services should always have user goals as their ultimate purpose. In today's technological context, there are many more alternative options for meeting user needs more economically, efficiently, and timely than a large body of collection. Without user-centered relationship management of collections, libraries cannot design processes to meet those demands. Designing collection processes also requires libraries to recognize staff and organizational capacity, although matching capacity with the forecasted collection needs is often not a part of collection development work.

As capacity and collection planning both rely on the knowledge of collection requirements, user relationship management has to be a prerequisite for excellence in collection services. It will connect librarians and library staff to the real issues of their constituencies, and therefore address them in collection processes on a regular basis. The current model in which collection tasks continue to be organized does not offer either the possibility or agility to address large-scale collection needs. Bell (2014) pointed out that UX, at its core, is about understanding people and their needs and developing good experiences that meet and potentially exceed those needs. It can only serve users better if libraries start with the understanding of patrons and their needs.

COLLECTION ENRICHMENT AND PRESERVATION

In the author's opinion, collection enrichment and preservation are also core expertise areas of libraries. Curation is an all-encompassing term that describes management of collections throughout their life cycle, while enrichment and preservation are two distinctive stages of that life cycle, which libraries often outsource. Again, there is a difference between a function and the types of tasks representing that function. Regardless of the fact that some types of jobs may be outsourced, core expertise should stay with the libraries. Collection enrichment and preservation functions are unique to libraries and aid in achieving their mission in providing long-term access to knowledge. Collection enrichment can be seen as any value-added enhancement to the content acquired from vendors and may include such examples, as classification and inventory achieved through bibliographic and metadata records. Those enrichments offer users better discovery experience and access to the collections, and afford libraries or a network of libraries vital information to make important decisions on weeding, conservation, and preservation of resources. The enrichments also gain from the knowledge standards, protocols, and system that libraries created. Again, taking collection description as an example, libraries hold expertise and know-how of classifying, describing, sorting, and relating individual items that are part of a browsable and searchable body of knowledge. Over time, library communities standardized collection description by streamlining and simplifying the discovery experience across institutions. No matter where scholars or researchers relocate and do their work, they are greeted with consistent experience. So to summarize, organization of knowledge is core expertise of libraries. It is unfortunate that libraries became so absorbed and immersed in development of their own standards that their practices failed to recognize the need to keep evolving with contextual and environmental changes taking place around them. There are lost opportunities for training library professionals to adopt new standards and practices that are much suitable for the digital age. Nonetheless, libraries have always been community and industry leaders in regulating collection descriptions for better discovery and long-term preservation and should remain such. Libraries have built much expertise and invested resources in knowledge management. Collectively, it is a significant knowledge workforce that remains very much relevant today. Libraries, however, need to keep sharpening their skills to develop best practices and embrace emerging practices. For example, skills developed by technical services employees have much overlap with those of the skills in web and information technology areas. Technical services will need to learn skills that will better integrate their function with the Internet-based strategies.

Libraries, of course, have options to completely outsource the supporting functions of curation, enrichment, and preservation to third parties, but the author believes that it would be a costly approach. Seceding their role of leadership and know-how, libraries will no longer be able to contribute to discovery experience. Some of the current sub-optimal discovery experiences are found in examples of vendor-controlled platforms, technologies, algorithms, and applications of various metadata practices. Perhaps libraries may consider outsourcing work if capacity is an issue; however, they should never consider letting go of the ownership of the collection enrichment and preservation. Libraries need to realize that there are no other entities in the entire knowledge ecosystem that have developed such a depth of expertise in collection enrichment and preservation. Libraries should advocate and champion the standardization of collection description practices and preservation practices across the collection services supply chain, a task that means working with vendors to unify the processes of library supplies for the benefit of users and long-term care of universal knowledge.

RECOMMENDATIONS FOR OUTSOURCING

Libraries should consider outsourcing based on their capacity constraints. Before outsourcing, libraries should carefully evaluate their operations. Without the understanding of their capacity limitations and their performance, it would be impossible to measure outsourced performances. For any outsourced projects, libraries will need to allocate resources to manage planning and quality assurance (QA) components. Often, the outsourced work, such as digitization or cataloging, is finished without any QA at the receiving end, which creates rework and impacts the discoverability of those outsourced resources. When libraries outsource driven by their capacity needs, outsourcing can be an effective strategy. Libraries may engage their user communities and vendors in completing tasks along the supply chain. Later in this book the author will discuss some possible innovations along the collection services supply chain related to outsourcing.

REFERENCES

Bell, S. J. (2014). Staying true to the core: Designing the future academic library experience. *Portal: Libraries and the Academy*, *14*(3), 369–382. http://dx.doi.org/10.1353/pla.2014.0021.

Davis, P. (July 22, 2014). *Libraries receive shrinking share of university expenditures*. Scholarly Kitchen. Retrieved from: https://scholarlykitchen.sspnet.org/2014/07/22/libraries-receive-shrinking-share/.

Esson, R., Stevenson, A., Gildea, M., & Roberts, S. (2012). Library services for the future: Engaging with our customers to determine wants and needs. *Library Management*, *33*(8/9), 469–478. http://dx.doi.org/10.1108/01435121211279830.

CHAPTER 12

Unique to Libraries

COLLABORATION IS THE KEY

Supply chain implementation requires close partnership that begins within libraries and across departments which own parts of the supply chain processes, such as plan, source, make, deliver, and return. Internal collaboration sometimes can be difficult not because of unwillingness of the parties involved, but because of shortcomings of most organizational structures. In the later part of the book, the author will introduce some ideas that will foster internal collaboration.

The implementation of the collection services supply chain also requires libraries to partner with vendors and patrons. It is a challenge to describe the relationship between libraries and their vendors as partners or collaborators because no such precedent exists. Libraries and vendors separately manage their respective end-to-end processes. The interface between them is routinely based on a sell–buy transactions model or occasionally gets amended driven by new product features. There have been instances where libraries and vendors partnered to create new products jointly; however, those endeavors were often aimed at designing solutions that better aligned with libraries' processes and increased market shares for the vendors. As far as end users as concerned, libraries seldom consider them as partners; the interaction between them is limited to a one-way activity of downloading or delivery of content. There have also been examples of libraries harnessing user communities to process collections; however, such efforts have not been systematically and broadly adopted by the industry.

The objective of the supply chain partnership designed for collection services must start with the end result of maximizing user productivity, instead of focusing on optimization of library or vendor productivity. If all improvements or optimizations were to gear toward user productivity, it is likely that occasionally library or vendor productivities will come in conflict with that of users; so, the bottom line is that intentional focus on user productivity, not that of the libraries or vendors, will bring the desirably high quality of services to users.

Supply Chain Management for Collection Services of Academic Libraries
ISBN 978-0-08-102031-9
http://dx.doi.org/10.1016/B978-0-08-102031-9.00012-8

Libraries should recognize that the internal collaboration among departments is the first step to delivering better services. In the current environment defined by proliferation of the supply and demand of knowledge versus scarce internal library resources, and personal enablement by the web versus limited processing capacity, libraries should become challenged to take internal cooperation to the next level of more seamless integration of their collection service processes with those of their suppliers and patrons. The partnership between vendors, libraries, and users offers opportunities for all entities to discover supply chain issues and thus explore solutions that may pool resources together and leverage each particular expertise under new circumstances.

ECONOMY OF SCALE AMONG LIBRARIES

Perhaps part of the library culture that may predestine a successful supply chain implementation is the ability to collaborate. Libraries formed alliances with one another and continue to do so to achieve a higher level of user productivity. Collections and services surrounding them are often the means through which libraries collaborate. The interlibrary loan services, shared print programs, shared preservation, shared facilities, shared cataloging, and shared acquisitions all leverage economy of scale and collective power obtained from like-minded regional, national, and international institutions, consortia, and associations. The collaboration among libraries serves local and networked patrons very well, and thus apparently accomplishes what libraries have been working toward. The industry-wide model of partnership creates opportunities to capitalize on individual institution's resources, supply chain processes, and capacity of the overlapping expertise to create much more sustainable, efficient, and economical collection services than any one of the individual members can do alone. The partnership proved to be productive in the past when library community was facing the same challenges and operated with the same set of constraints: production of physical items, their circulation and usage, and time and space limitations of their transportation. The set of constraints defined the environment for the entire knowledge industry back then and was characterized by slower pace of production and dissemination of knowledge and higher technological and fiscal thresholds for outsiders to enter into the knowledge industry. In contrast, today's patron expectations are driven by the desire to have access to requested items almost immediately.

If libraries were still operating in the era of predominantly physical collections, the previously established level of partnership among libraries would suffice; however, libraries have shifted and are now operating in a new space of digital collections with different sets of challenges. The author believes that under the current circumstances operating in a siloed environment and attempting to overcome all the hurdles is no longer an option for a single library or even the entire library community without expanding existing collaboration realms. User expectations for collections are rapidly evolving, heavily favoring convenience and instant access to needed content. As the entry to publishing market and sharing of knowledge has become easier, more nontraditional competitors are joining the sector and creating substantial amount of content on a daily basis. Library resources and capacity are getting even more scarce, struggling to handle the increased body of knowledge. Libraries have no option but to expand beyond existing partnerships within their own network.

BEYOND BUY–SELL TRANSACTIONS

The same set of emerging challenges also applies to publishers and library vendors and technology suppliers. Library vendors have also been shifting from a controlled content production environment to the new realm of scholarly communication. Library vendors can no longer afford to operate strictly within their industry networks as they look for ways to conquer new challenges. Just like libraries, vendors may choose to ally with others in their sector or take more market shares through mergers and acquisitions, but the author believes that the new context dictates a different type of partnership, a partnership of knowledge supply chain integration with their suppliers (in this case, could be authors) and patrons (in this case, libraries).

The fundamental goal of the new partnership model is to reach scalability for production of better service results by sharing and capitalizing on unique and duplicated expertise areas. The collaboration along the supply chain processes can eliminate waste of resources and thus lower costs that individual players would otherwise incur working alone. Successful collaboration can also increase capacity of all parties involved to embrace new technology and innovations. The partnership should optimize collection processes with an ultimate goal of realizing "Make-to-User" just-in-time collection services.

As partner industries continue to take advantage of the benefits of their own networks, the new partnership will afford more options for harnessing core competencies, expertise, and processes. As partners, they should reimagine what such new collaboration model can do to solve collection services issues. Currently, the vendor–libraries relationship remains mainly in a form of a seller–buyer relationship. Libraries and vendors should investigate ways for innovation based on the concept of the extended supply chain.

This new partnership model should promote shared purposes, goals, and rewards. It may appear counterintuitive to form a mutually beneficial partnership because of the "struggle" between libraries and vendors on commercial pricing models and open publishing models; however, both groups should have end users in mind and work toward meeting their needs for knowledge. If by putting the competition aside, libraries and vendors can reap greater benefits from the efficiency of transfer of information from vendors to libraries, and eventually to users by

1. Predicting user needs better by utilizing reversed information flow from end users to libraries and vendors that will help both parties better understand user demands;
2. Streamlining the supply chain processes for just-in-time delivery of content directly to users.

Vendors and libraries must maintain open lines of communication and must cultivate mutual trust through their supply chain integration. Vendors have to believe libraries' desire to get the best in class knowledge available to their patrons, and, in return, libraries need to trust that vendors structure their operations around the notion of having users' best interest at the core of their business principles. If libraries and vendors move jointly and within such close partnership, not only can they together develop processes and workflows to present existing content, but also more efficiently develop new content, services, and products to satisfy the growing market demands. The above benefits simply cannot be realized as part of the current seller–buyer relationship.

When forming the new strategic partnership, libraries and vendors perhaps should aim for small wins to gain more bilateral trust over time. The current relationship has been very fragile and delicate, pressured by library budgetary reality and seller profit expectations. Only when vendors and libraries realize that they are in the same ecosystem and make a joint decision to form a better working relationship, will they be able to understand system issues, so the ecosystem works and functions well. Libraries and vendors must approach collection issues from the angle of shared purposes, goals, and potential rewards, and thus focus on long-term sustainable benefits to each party. Any short-term fixes will only push the relationship further apart and lead both sides to eventual damages and win–lose situations, which is detrimental for end users that both groups serve.

Libraries should continue to advocate for their users with understanding that vendors also have a set of specific challenges and constraints they face and that they need to sustain their business in order to serve end users. At the same time, libraries' advocacy efforts should challenge providers' current practices and help them perform better for end users' sake. At this point, readers may wonder whether a real collaboration between libraries and vendors is even possible. The truth is that not every entity sees collaboration and partnership to be in its best interest. Only if organizations understand their ultimate goal of serving end users, would a real and meaningful partnership become a possibility. Successful partnerships should center around shared information on such issues as capacity, resources, user needs, obstacles, and expertise. The author recommends libraries and vendors set their partnership relationship straight from the beginning:

1. Develop an agreement which establishes purposes and expectations for the partnership;
2. Create a shared plan which outlines goals, milestones, strategies, as well as steps to execute the plan;
3. Create an understanding of emerging service and pricing models.

The above steps should establish an understanding of minimization of pricing disruption prevalent in the current model with the acknowledgment of service models and pricing changes in the future. The continuation of operation is critical because it ensures the profits required to sustain vendor business; meanwhile, the stability of the supplier would guarantee libraries could obtain flexibility to work out a new service model that offers a sustainable pricing structure. The agreement will serve as a base to streamlining collection workflows, sharing expertise and responsibilities along the supply chain. Later part of the book will give examples on some of the supply chain innovations centered around the topic of library–vendor partnership.

FRONT, MIDDLE, OR END OF THE CHAIN

The same set of new challenges also stipulates libraries to partner with patrons. Libraries must recognize some of the new roles in the area of scholarly communication that users can play in the current technological context. Most of the time, library patrons only play one role, that of consumers. Under the current climate, since library users can be patrons of many other information providers, why should they be loyal to libraries? The entire society is shifting from a closed knowledge system approach to a much more open space in the area of scholarly communications. Libraries must play a more proactive role to engage their users. Also, by acknowledging the fact that libraries are not sufficiently resourced to solve their collection issues alone in today's environment, libraries should partner with as many potential interested parties and as widely as possible. It is easier for libraries to collaborate with vendors because such relationships already exist, but it can be harder to work directly with patrons. Historically libraries have been playing a role of experts in knowledge selection and delivery. Partnering with user population will require an extreme mindset shift from established well-defined lines of authority to a new form of yet unknown relationship. It might be difficult for libraries to adjust to and perform in the new types of relationships, particularly with patrons in a role of peers and collaborators.

To make the new partnership work, libraries must also understand the shared purposes, goals, and rewards of the partnership they are creating. In this new relationship libraries need to respect the set of expertise that patrons bring to the table. Patrons can play a variety of roles in collaboration with librarians. First of all, librarians should recognize users' expertise in their disciplines. As libraries increasingly have difficulties in filling every subject liaison position due to the shortage in funding, they should seriously consider involving users in collection selection. It is a fact that patrons, particularly those who hold faculty status, have to know what they need for their research and scholarship in a university setting. Since librarians have more duties than ever before, partnership with users in collection selection will free up valuable time for other critical work librarians can focus on. As libraries leverage patrons for collection selection work, the domino effect on resource savings will reach all financial and staff time resources which currently often remain locked under the budgetary line dedicated to collection preparation. The above statement comes from the observations and statistical analysis of library circulation data that demonstrate patrons used fewer items than libraries ordered for them. The patrons' involvement in collection selection may translate into work for libraries to make a catalog of recent and updated publications and create processes that handle just-in-time orders and delivery of content. As the Part Two—Library Supply Chain Model of the book recommended the just-in-time "Make-to-User" processes should not eliminate the needs for "Make-to-Stack" processes, in which libraries should continue to acquire must-have and special collections for the campus; however, the must-have should be determined through consultations with patrons.

Patrons can also participate in activities of Make processes of collection services. It is obvious that patrons can either be authors of the types of content that libraries acquire or peers of the authors of the type of content that libraries acquire. To keep patrons in the loop of what is available and more so of what is upcoming will create opportunities for them to engage with basic curatorial work. At the very least, patrons may participate by adding important access points for particular collection items, and they can certainly do even more if the system, technology, and design come together offering an intuitive user experience. Libraries should approach such partnership opportunity at the very first step of knowledge creation and engage patrons in the process of shepherding their scholarship and research through the complete life cycle. A real example is the author's current institution, where the libraries are integrating user research workflows with the digital repository (Open Science Framework for Institutions and CurateND Integration project). In such model, faculty and their research teams will take care of many of the "Make" processes, with necessary assistance of librarians. As researchers move along with their research, data will be properly archived and labeled and eventually pushed into the repository (CurateND) for discovery and preservation. The new partnership opens many possibilities for libraries to engage in scholarly communication conversations, particularly in those cases that are scholarly valuable but unprofitable.

The above example may show some similarity with the open access initiative, where libraries move into publishing world; however, working with end users should not be in conflict with working with vendors, even though at times it may feel like that. Libraries should be the intermediaries that bring content providers and end users together to solve research, learning, and teaching productivity issues. The conversations should aim at enabling expertise-based partnership. Again, there should be an established respect and reciprocal understanding of what vendors bring, what libraries bring, and what users bring to the process. Ideally, where applicable, the partnership should engage all the three entities through integration of separate supply chains—the publishing supply chain, the collection supply chain, and the user knowledge supply chain. In other words, each entity should move up and down the extended supply chain freely and be able to contribute their expertise. Later part of the book will discuss this idea in more detail through examples of supply chain innovations.

CHAPTER 13

Disruptive Innovations of the Collection Services Supply Chain

OPEN ACCESS COLLECTION SUPPLIERS AND PARTNERS

Open Access (OA) is a relatively new publishing method that is highly disruptive to the commercial publishing model. Based on Zuccala (2010), OA is a scholarly communication movement that was developed by scholars to increase the impact of future scientific research and create a cost-effective publication system. OA publishing has taken many forms and spread beyond scientific disciplines at this point in time. The traditional publishing model is a paid system, where the cost is absorbed by institutions, such as libraries, or individual authors. Part of profit goes back into the system, sustaining the model, enhancing technologies, supporting staffing, building infrastructure, and managing visibility and marketing. Morgan, Campbell, and Teleen (2012) in their article provided a very thorough example of journal subscription paid model and the various costs that publishers need to recover from annual subscription fees. OA challenges the sustainability of the commercial model because OA promotes either no payment or one-time fee to make publications freely available to readers. Without the annual subscription fees or the subsequent sales of books, publishers will not be able to continue fueling their operating systems.

Some may argue that the commercial model is either obsolete or it hinders the promotion of scholarly communication. The author agrees that the increasing costs, particularly the annual subscription costs, and the lack of real-time ability in dissemination of publications jolts the stability of the model; however, OA also requires resources to maintain the process. The truth is that a one-time fee just cannot sustain the system in perpetuity unless the fee is astronomically high. Since the publishing field is big enough to accommodate more than one model, OA or commercial, the author suggests investigating supply chain integration in Chapter 14 that publishers and libraries may collaborate on with a goal of increasing efficiency of knowledge supply. Cost is a central issue of the supply chain integration. No matter if it is OA or commercial, someone has to pay to support and sustain the model, which could be the common ground for cooperation between

Supply Chain Management for Collection Services of Academic Libraries
ISBN 978-0-08-102031-9
http://dx.doi.org/10.1016/B978-0-08-102031-9.00013-X

publishers and vendors. As libraries and publishers eliminate waste in their processes and improve the forecasting of the user knowledge needs, chances to create a sustainable funding model is highly feasible. The author believes that both OA and commercial models will come up with an economical and efficient solution to promote scholarly communication as publishers and libraries segment publishing needs and share expertise and infrastructure. OA is a disruptive model that offers opportunities for both publishers and libraries to realign their values along the collection services supply chain by adapting and revolutionizing their current processes.

Patron-driven acquisition (PDA) and demand-driven acquisition (DDA) set the precedence for a use-based acquisition model. Although some publishers are greatly concerned about the impact of such models on their current sales income, in some way, it is a reality that publishers are beginning to come to terms with. It is evident that libraries have started a movement to change their conventional acquisitions models. More libraries will recognize the benefits of the use-based model when they notice the savings they generate from such programs that could then be invested toward other growing demands. PDA and DDA may come as an evolution of how vendors and libraries are coping with the reality of library purchasing power; however, the underlying use-based philosophy is revolutionary from the point of view of the long-standing practice of collection building. While PDA is primarily implemented with ebook services, the use-based collecting method could be expanded to print and serials. It is just a matter of pre-populating records into library search environments. The use-based capabilities are abundantly available outside of library vendor supplies. Libraries have broader selections when it comes to working with other vendors to accomplish this goal, which is disruptive to current library vendors. More potential distributors can enter this market or libraries can actively solicit services from others outside their traditional supply chain. The threshold to get into the market is very low since distributors and publishers only need to figure out a way to provide libraries with their catalogs. Generally, in the current environment, the channel is created through application programming interface to library systems and at the backend it integrates with acquisition processes. Tied directly to the "Make-to-User" strategy introduced in Part two—Library Supply Chain Model of the book, there are many opportunities for libraries and vendors to create supply chain partnerships in which they can streamline their acquisitions and collection delivery processes from orders to direct delivery of their publications to users, regardless if the format is physical or digital.

For the collection services supply chain, the process integration or innovation is key, which will be a significant leap from what libraries are currently buying from some nontraditional wholesale sources, such as Amazon. Many libraries have used Amazon to order use-on-demand books, but the process is merely an addition to their existing acquisition workflows. Amazon is another source or channel, just like ILL and some firm order or rush order vendors, which libraries leverage for quick delivery service, but the process itself is not an end-to-end optimized one that can initiate directly from patrons and go all the way back to Amazon. Such streamlined process is disruptive for the current acquisitions workflow, but it should not be seen as a threat to libraries' vendors. Lately, many vendors have invested in and kept upgrading their "dedicated channels" to content aggregators, as well as libraries; at the same time, many publishers have supplied Amazon with their academic titles as an additional sale channel. The opportunity for publishers would be to leverage the strong logistic capabilities of Amazon and other similar vendors to provide collections to libraries. The author sees no need to develop duplicate channels, which, in any case, would not even be one of core competencies of publishers. In the author's opinion, such channel integration is within reach, only a few conversations away. Historically, library vendors and libraries do not come together to discuss issues of the extended supply chain, yet based on the "Make-to-User" model, there are tremendous opportunities to send online collections directly to users. One good example of the supply chain innovation or integration is the workflow developed by OverDrive that enables patrons to download ebooks and return them from their Kindles, which demonstrated possibilities for the academic market.

SELF-PUBLISHING CONTENT

The market for content services has been changing dramatically, enabled by the Internet. The knowledge industry has seen many shake-ups in the form of mergers and bankruptcies in the past 3 to 5 years. The incumbents of the publishing industry are forming larger corporations, achieving scalability by taking advantages of joined expertise, capabilities, and capacities. Mergers are a normal phase of the market for vendors to gain market shares, capitalize on the scale, and push for innovations. There is some level of fear and anxiety from the libraries; seeing the shake-ups in the industry could mean monopoly and henceforth more dramatic price increases for acquiring content. It is a possible although an unlikely scenario because it will not be a sustainable one either for vendors or libraries. Another possible scenario is that the price might be leveled off because the scalability is achieved from those mergers. Libraries should view this as a chance to play a more influential role in shaping up the new market by collaboration with vendors. Supply chain may offer a shared vision of both vendors and libraries as it could eliminate waste, achieve process efficiencies, and thus restructure the current pricing models. The current pricing model has to be counted in the total cost of knowledge production, including all labor, equipment, infrastructure, transportation, and marketing expenses. Supply chain conversations can offer libraries and vendors a chance to reexamine the flows of processes, information, as well as content in their extended chain, and thus find areas for improvement which can trigger opportunities in the current pricing model. Collaboration is a much better approach in regards to sustainability and long-term relationship between publishers and libraries.

An example of a more disruptive innovation for both libraries and vendors is content created by nontraditional authorities. The previous section of the chapter discussed OA. Libraries and vendors alike are figuring out ways to incorporate OA in their existing business models. Many libraries have fully integrated OA journals and monographs into their discovery platforms. This section will not discuss OA in more detail, except it will mention that OA will stay and continue to evolve. Earlier in the chapter, the author already recommended how libraries and vendors could approach OA; so this part will focus on individual publishing or indie publishing. The entire knowledge industry has not decided yet on a way to deal with personal publishing. With the increase in importance of digital publishing, more than half of the titles published in the United States are self-published (Holley, 2015). The initial reaction of our industry to this new style was the denial of its value. Readers will remember rejection of Wikipedia and folksonomy as an information source and a classification system within the library community. It happened because it was something very different from the industry's naturally authoritative system. Today, self-publishing has spread beyond Wikipedia to scholarly blog posts (a form of journals and their articles), monographs, and even special collections and archives. The quality of those self-published materials is still a primary concern for many library professionals; however, when a comparison of the authoritative and self-publishing channels is made, both systems exhibit surprising similarities. In the author's opinion, a self-publishing channel, empowered by the Internet, provides opportunities to talented writers and experts who were not given a "voice" or a "publishing" platform through the traditional authoritative channel. Comparing some Wikipedia entries with those that appear in traditionally published encyclopedias, the level of scholarly caliber is comparable. Many of those Wikipedia entries were composed and edited by well-educated people, and the value provided through the self-publishing system is analogous to the peer-review process. Granted, all self-publishing efforts are not going through the equal level quality control filter of the crowdsourced editorial efforts and critics, yet the infrastructure of the traditional authoritative channel just cannot accommodate all publishing demands at a pace that the society of today requires. So the idea is that both publishing channels need to be recalibrated to some point, which is a strategic area that libraries and vendors need to consider jointly. As far as processes, libraries and vendors need to decide on the methods that would allow facilitation of getting cream of the crop in self-publishing and funnel it through their extended chain to users. In other words, there needs to be a process that could surface good candidates for traditional publishing, while at the same time prevent vanity publishing endeavors. The author firmly believes in the value of peer-review and library collection development processes, but in today's world, vendors and libraries must look to expand the scope of "peer" and "librarian" pipelines for the ultimate benefits of users. The author believes that publishers and libraries do not lose their authority when they open up their industry processes to embrace self-publishing; on the contrary, such move will continue contributions to their authority on quality. The bottom line is that there is much untapped talent that remains outside the walls of traditionally established system that could join in and add great value to libraries' and publishers' endeavors.

It is certainly a part of knowledge creation process that professionals do not know how to deal with, which emphasizes the importance of libraries and vendors working together to incorporate them into their supply chain, notably source and make processes of their publication to collection services continuum. Source is represented by getting self-published content into acquisitions processes, and make is to enrich self-published content for better discovery. Technology and Internet platform will have to play a critical role in integrating self-publishing method into the current knowledge system. The established commercial or OA systems have limited resources to address self-publishing. Libraries and vendors will need to open their internal classification systems to the world, further adopt semantic linking technologies and standards, as well as enable crowdsourcing for make processes and the evaluation of the quality of publications. As it stands, current discovery and delivery platforms are still closed to self-publishing. One innovative example of bridging the gap that the author can share is IngramSpark, a support service for self-publishing authors. IngramSpark has mechanisms in place to support sales to libraries and bookstores (Cutler, 2015). The following part of the chapter will discuss bibliographic services and crowdsourcing.

OPEN BIBLIOGRAPHIC DATA SERVICES

Chapter 12 discussed the enablement of sharing bibliographic data along the extended supply chain. There are cataloging workflow steps that should be rethought as many of the work can rely on publishing metadata coming directly from publishers; at the same time, as collections go through their life cycles from access to preservation, or from general to special, libraries can participate by enriching publishers' metadata with preservation, as well as holding data. Chapter 14 will describe opportunities for upgrade of the underlying technology supporting metadata and bibliographic records, suggesting that a move is made from the machine readable format to machine comprehensible style by adoption of semantic linking techniques. Some of the benefits of the switch include its openness, expandability, scalability, and accuracy in content search and discovery. Libraries and vendors should consider the above for their supply chain processes, but this section will discuss the possibility of opening their bibliographic data services to content creators. In reality, those bibliographic standards have already been open to the world, but the "open" in the context of this chapter is the process that requires content creators to link their individual scholarly works using established standards. Having such a process in place has some indisputable advantages. First of all, it creates metadata for discovery from the source, which is related to the implementation of the extended supply chain. To have authors contribute metadata for their own works pushes the chain even further back. Many publishers already request keywords from contributors, but at this stage, publisher metadata workflows are not open to authors directly. One of the possible reasons for that could be the technicality of publisher metadata, but the author believes that WYSIWYG (What You See Is What You Get) tool can be created for authors and a codec can be put in place for generating publisher metadata automatically. Self-publishing authors definitely need this type of support to make their work organized better and incorporated into a formal classification system. Much self-published content lacks visibility and recognition due to absent metadata support. This approach alone will not address quality issues that accompany the self-publishing technique, but by integrating it into a formal process it will most definitely help increase the value librarians and publishers bring to the exercise of filtering the best content, while at the same time also allowing users to judge the quality of publications. A good analogy is the NetFlix movies rating and Amazon book rating mechanisms, in which viewers and readers contribute to the product and content quality evaluation.

PATRONS IN NEW ROLES AS COWORKERS

Supply chain methodologies would elevate patrons from their single role of content consumers. Chapter 15 will discuss the value of starting with patrons as content consumers first when designing a collection services supply chain. This chapter will present some other significant roles that patrons can play and that will contribute to the improvement of scholarly communication. Knowledge industry institutions need to realize that a percentage of library users will create and contribute new content at some point. Libraries provide important educational and research materials in support of their new writing projects; in return, what users publish will enhance library collections. Patron's potential role as a writer is a well-understood and accepted one, so, if libraries and vendors acknowledge existence of their users' authorship, they should also recognize their expertise in the subject areas they create. The entire knowledge supply chain at this moment only taps into patrons' expertise in creative writing, but not in other possible areas for collaboration, such as providing reviews, curating, and organizing content. One emerging role the author sees patrons playing is being peers to library professionals. Granted, not all patrons would like to play that role, and, certainly, not all patrons who would like to step into this role have the necessary credentials to fulfill it, yet the open attitude of considering them as collaborators is absolutely critical to navigating today's knowledge management challenges.

In the broader context of the supply chain, publishers and libraries may choose to engage their patrons further than just in the knowledge creation phase. The supply chain stages such as make can particularly benefit from patron's knowledge in their respective disciplines. Make process is the stage at which enrichment to collections takes place and includes such processes as metadata, conservation, and preservation. One of the apparent functions that patrons can contribute to is the metadata enrichment. In that setting, patrons play a role of curators, adding meaningful metadata either for identification, description, preservation, or scholarly evaluation. There are many success stories about libraries' crowdsourced projects where institutions leveraged their patrons' communities to enrich collection metadata (Ashenfelder, 2015; Corney et al., 2010; Patrick Baggett, Bridges, Wise, Tanner, & Mezick, 2014). The extended collection services supply chain should take advantage of the crowdsourcing movement and more specifically apply it toward collection curation. Publishers can begin by assigning particular rights to authors that would empower them to generate metadata for their own publications. As the information about their publications would move downstream, publishers and libraries can add more metadata to supplement and enrich the existing descriptions. Authors should be kept apprised of the enrichments applied to their works throughout their life cycle. As author communities begin to grow within the supply chain, publishers may also notify the communities about their peers' publications, either soon-to-be published or already acquired by libraries. This type of collaboration is dynamic, live, and real-time, which not only increases patrons' productivity in scholarship and research, but also offers a chance to contribute back as peer curators, adding more metadata, writing reviews, etc. User participation could be very useful for publishers as it will inform their planning decisions and forecast readership demands. It is also invaluable for libraries as they may receive user generated information about their collections and apply it toward forming plans from acquisitions to preservation. Finally, it is helpful to users since they proactively engage with scholarly communications. Even if patrons only contribute to metadata creation, the process could be streamlined in such a way that many possible duplicated steps would be eliminated, making the process much more scalable and economical than the currently existing discrete metadata processes used by library and publisher industries. If publishers and libraries are bold enough, they may choose to include all faculty and graduate students in their supply chain and not limit their roles only to those of authors. Supply chain technology should consider focusing on enabling such massive collaboration effort among publishers, libraries, and knowledge creators. The potential level of information that publishers and libraries can generate from this level of cooperation goes way beyond what the current supply chain is able to provide.

COLLABORATIVE STORAGE SERVICES

One last topic this chapter will briefly mention is possible shared storage services. This is an area that has been separately managed by publishers and libraries. Libraries have formed coalitions on storage services. Publishers probably have been looking into 3PLs (third party logistics) to manage their inventories. The author sees that there is an opportunity for collaboration between publishers and libraries. Storage probably can be classified as inventory space, open collection space, and preservation quality space. As inventory control is an essential concept of the supply chain management, this book introduced it as part of collection services supply chain. The notion of inventory control should shift library collection practices from blanket purchases to demand-based acquisitions. Currently, from the perspective of the supply chain, many of library open collection and preservation spaces are utilized to house inventory in a form of collections. In the future, libraries should consider leveraging vendor's storage facilities as inventory space instead of buying unused or low used collections to occupy precious physical spaces. By doing so, it will lower the level of frustration about space shortage and total costs to maintain a large body of uncirculated content. The inventory should apply to both print and digital publications. The make-to-user and make-to-stack models are specifically designed to treat unused content as an inventory of publishers and not as collections of libraries.

On the other hand, publishers should leverage library collection and preservation spaces as their preservation spaces, both for print and digital formats. Preservation is part of library mission and part of library core competencies. As the author earlier in this chapter proposed an idea of sharing a metadata standard across the extended collection services supply chain, he also suggests that publishers should gain holding information from libraries about the availability and status of their publications, which, in turn, should inform publishers on their inventory and reprint decisions. Furthermore, publishers, technology vendors, and libraries should come together to solve the storage issues. The author assumes that there are plenty of storage spaces, but they are not equally distributed or utilized appropriately. If publishers and libraries can share their current storage spaces, this endeavor will solve many of the supply chain inefficiencies.

This chapter only listed some areas and provided a handful of examples where libraries and vendors could innovate collaboratively along their extended collection services supply chain. These areas and examples are not meant to be all-encompassing. As libraries and vendors join forces to tackle their supply chain issues, the author is confident that they will find other opportunities to improve their processes for the benefit of end users.

REFERENCES

Ashenfelder, M. (September 16, 2015). In *Embracing crowdsourcing* Retrieved from: http://blogs.loc.gov/thesignal/2015/09/cultural-institutions-embrace-crowdsourcing/.

Corney, J., Lynn, A., Torres, C., Di Maio, P., Regli, W., Forbes, G., et al. (2010). Towards crowdsourcing translation tasks in library cataloguing, a pilot study. In *Digital ecosystems and technologies (DEST), 2010 4th IEEE international conference* (pp. 572–577).

Cutler, R. (2015). Ingram and independent publishing. In R. Holley (Ed.), *Self-publishing and collection development: Opportunities and challenges for libraries (Charleston insights in library, archival, and information sciences)* (pp. 83–101). West Lafayette: Purdue University Press.

Holley, R. (2015). *Self-publishing and collection development: Opportunities and challenges for libraries (Charleston insights in library, archival, and information sciences)*. West Lafayette, Indiana: Purdue University Press.

Morgan, C., Campbell, B., & Teleen, T. (2012). The role of the academic journal publisher and open access publishing models. *International Studies Perspectives*, *13*(3), 228–234.

Patrick Baggett, M., Bridges, A., Wise, K., Tanner, S., & Mezick, J. (2014). Populating the wilderness: Crowdsourcing database of the smokies. *Library Hi Tech*, *32*(2), 249.

Zuccala, A. (2010). Open access and civic scientific information literacy. *Information Research: An International Electronic Journal*, *15*(1).

CHAPTER 14

Emerging Standards, Practices, and Technologies

Previous sections of this book discuss supply chain innovations and collaboration opportunities. This chapter will cover some emerging standards, practices, and technologies that a supply chain cooperation between libraries and vendors may focus on along their Extended Collection Services Supply Chain. External relationships are difficult to create and maintain, in general, but in particular, between libraries and providers in this economic climate. Libraries find themselves feeling squeezed budgetarily; yet, vendor's content prices keep steadily rising. There are no adequate solutions that address the tension. Supply Chain might offer some options for libraries and their vendors to rebuild their relationships, explore possibilities for new services, and improve existing services. Ultimately, the external partnership must aim at creating benefits for end users. Such a relationship will significantly eliminate waste in current operations, improving the profitability of vendors, and lowering the budgetary stress of libraries. Supply Chain is not an end-all process improvement methodology; it is a strategy to improve competitiveness and overall operation efficiency. The partnership may also boost the speed of recognizing innovations and new services and thus bolster propensity to strategically allocate limited resources to critical growth areas. The author recommends libraries and vendors build processes and systems that leverage:

- Core competencies of vendors and libraries
- Internet standards
- Streamlining of collection processes for the Make-to-User strategy

VENDOR CORE COMPETENCIES

Definition of core competence is a strategic conversation that libraries and vendors should conduct first, as it decides on what do and what not to do as part of collection services.

The concept of core competency is not unfamiliar to libraries and vendors. Core competencies are unique skills. The concept was developed by Hamel and Prahalad (1990) for the purpose of increasing

Supply Chain Management for Collection Services of Academic Libraries
ISBN 978-0-08-102031-9
http://dx.doi.org/10.1016/B978-0-08-102031-9.00014-1

competitiveness. If organizations perform outside their core expertise, they divert valuable resources, efforts, and energies to activities that do not return any value. According to Hamel and Prahalad, competencies are the roots of competitiveness. Hamel and Prahalad also define core product and end product. If collections are libraries' end services, the following core services will be part of the end services: content itself, search interface of the content, curation of the content, metadata describing the content, and preservation of the content. Along those lines, the author will focus the rest of the chapter on those competencies suggested for use by vendors and libraries.

Chapter 11 already discussed the core competencies and core products of libraries. What are the core competencies of vendors that support collection services? In the author's view, core expertise of publishers, system and technology providers, and libraries should follow their essential functions in the knowledge industry, as suggested below:

- Publishers—content identification, packaging, and promotion
- Technology vendors—enabling content process
- Libraries—content curation, enrichment, preservation

These days the lines between core and noncore have become increasingly blurred among vendors and libraries. Previously, suppliers and libraries brought in different sets of expertise and know-how to the industry. Publishers specialized in solicitation, review, selection, and packaging of content; technology vendors concentrated on publishing systems and library systems; libraries were exceptional at classifying, organizing, preserving, and providing access to content.

The above sets of skills are complementary, with each being practiced in a separate section of the overall Knowledge Supply Chain. Books and journals flow from publishers to libraries and then to users, enabled by vended systems. With time, libraries have moved into the publishing and systems businesses; publishers have been getting more efficient at content curation and classification and have been investing in technologies. The information industry has been seeing its members expand from their core functions into some other areas. This extension movement was perhaps driven by cost reasons and presence of various ambitions; however, supply chain calls for a different type of extension between libraries and their vendors. The author is not arguing that the way libraries and their suppliers have been expanding is not performing under the new set of circumstances; however, supply chain conversations would offer a new perspective on library and vendor relationships regarding their core competencies and core services.

Apparently, there are more demands on knowledge than the capacity of publishers, systems vendors, and libraries can handle individually. The publication rate is accelerating as more public and private investments are pouring into research and scholarly activities. Entities in the knowledge industry perhaps should consider segmenting and differentiating the demands by partnering on their core competencies.

It is evident that some publications are not and most likely will never become profitable or should ever target financial gain as their bottom line. Scholarship and research should not be evaluated based on profitability alone, while, at the same time commercial publishing channels continue to produce high-quality and reputable works. Perhaps publishers should collaborate on solving the dissemination of knowledge based on merit regardless of their profitability. Publishers should consider focusing on publications that sustain their business so that they can keep contributing to the scholarly communication; libraries should focus on distributing scholarly meaningful but not necessarily profitable works.

Publishers might consider providing publishing services to libraries, leveraging their core competencies and know-hows; libraries, on the other hand, may offer their curatorial and preservation expertise to publishers. The author sees the above potential partnership built on complementary expertise as a win–win situation for all entities, including technology vendors. The partnership requires systems that enable collaboration workflows. Such collaboration will release a host of resources that have been invested in overlapping efforts and thus provide opportunities to restructure the current pricing model and increase more visibility of important scholarship and research, which might not, otherwise, be profitable. Again, the above scenario is not a single possibility of a new type of partnership based on core competencies.

The author is also aware of publishers' desire to continue to expand their operations beyond their core competencies as means of growing their business and gaining market shares. The same desire will compel libraries to keep investing in ventures that would drive-down costs of operations and support scholarship and research. The author believes that such change is difficult to accept and implement and that it creates long-term sustainability issues. The author also realizes that it still comes short of addressing the issues of ever-expanding publishing demands and publishing equity. Construction of a partnership based on complementary expertise, on the other hand, will present possibilities for untapped creative solutions. Competency-based collaboration requires creation of a new supply chain.

This chapter offers some food for thought as libraries look to sculpt a new type of partnership between vendors and libraries and a new supply chain that leverages emerging standards, practices, and technologies.

SEARCH ENGINE INTEGRATION—VALUE OF SEARCH ENGINE OPTIMIZATION

The current state of collection services can be dramatically enhanced by integrating processes with the Internet as it would streamline the workflow across vendors, libraries, and end users. The Internet is clearly the starting place of many people seeking information. As Chapter 16 will discuss the importance of starting with end users first, the rationale for the integration of collection services workflow with the Web is also a core competency–based partnership.

Search algorithms are core competencies of search engine companies. Search is their core product, and some examples of their end products include website search, image search, video search, and scholarly content search. Those firms have invested heavily in improving their search algorithms based on the analysis of a large number of queries over time. It would be prudent to leverage those algorithms as libraries strategize collection discovery. At this time, vendors and libraries separately invest in search products for their niche search environments. Although library proprietary search technologies are useful for certain demographics, from a long-term perspective, search engines have far greater chances to succeed than closed technologies adopted and developed in libraries. For one of many reasons, the level of investment and resource input of search engines firms are not something that libraries or library vendors can sustain competition with.

Readers at this point may question the benefits of integration with search engines. Many libraries, consortia, and networks have made their catalogs, including special and archival collections, discoverable on the Internet. As the Internet was loaded with nonlibrary resources, which compete for user attention, library efforts seem not to be effective at directing more traffic to library collections. Integration with the Internet requires more than just pushing bibliographic records and collection Web Pages to the Web. Although pushing is the first step, libraries also must synthesize their processes with those of the Internet, meaning that libraries should see the Internet as the discovery platform and incorporate accepted Internet discovery standards and practices. More importantly, libraries should strive to contribute their curatorial competencies to the wider Web community.

Library community has well-established bibliographic standards for description, preservation, search, and discovery of their collections, widely used by its members. Compared with other industries, libraries have been successful at achieving and enjoying a network-level efficiency and scalability for sharing, curation, and preservation of collections for quite some time, which is greatly valuable to library users. The Internet gives new network possibilities and rescopes efficiency and scalability. Under the openness context of the Web, library standards became shifted to closed-door rules and protocols. Although crosswalk efforts have been made between the library and the Internet standards, libraries, in general, have not taken the time to understand the merit of the existing Internet standards and the benefits they would offer to the library community. The value of the Internet standards, in the author's opinion, is its evolutionary search logic used to relate and classify information. The Internet applies all possible means to make sense of content for users, for example, by keywords, full-text, credibility, reputation, and frequency of clicks. The author is confident that not many librarian professionals, perhaps technical services librarians being an exception, learned how search engine algorithms work and thus use their knowledge of search engines to enhance discovery experience for their patrons.

Library professionals should situate the discovery of library content in line with Search Engine Optimization (SEO). Other industries have already widely adopted SEO. It began to get some traction in libraries as well. SEO is a practice that pushes content to the top of the page of search results. Libraries have not yet switched their mind-sets from optimizing catalog and local discovery experience to that of the Internet. Just by allowing search engines to index their pages, libraries would make a move in the right direction; however, libraries need to learn to provide the "right" kind of information, the one that search engines need to recognize library content as primary resources. Libraries may have pushed their collections to the Internet, yet as long as the ways to promote them do not correlate with the requirements of the search algorithm, it would be challenging for the search engines to rank library resources high. The author has firsthand experience using SEO principles for oral history and archive websites. By creating keywords, metatags, and copyrighting the content based on search engine rules, the visibility of the pages on search result pages became greatly improved. The process for those exercises was rather manual, so automated or systematic processes should be put in place to facilitate SEO use in libraries.

The author believes that SEO is a skill that libraries need to master in the digital age to advance collection discovery. By adopting SEO, libraries will contribute their knowledge curation and management competencies at a global-level platform. Once visibility of library collection is made on the Internet, libraries should further integrate their login and delivery processes at the Internet level as well. The above were two examples of further integration at the process level.

The ultimate goal is to streamline the collection workflow as much as possible, so the use of the library collections can be as convenient as possible for patrons. The delivery processes should include both physical and digital delivery. Such level of integration is remarkably critical for the Internet age. Even back in 1952, Butler pointed out that "Separately a book stock and a catalog would be comparatively weak, but together they possess a unique cultural potency." Although Bulter referred to physical collections, the togetherness of metadata and content is why the Internet becomes a superior platform to that of library catalogs. HTML as a core language of the Internet brought metadata and content together and thus possesses a stronger cultural potency than library catalogs do. Currently, libraries provide metadata on the Internet, but they do not provide most of their content. Such practice puts libraries' content in an inferior status compared with other resources, even though library resources are often carefully selected primary resources. "Publications that are not instantly available in full-text will become kind of second-rate in a sense... become much less visible to the reader community" (Educom Review Staff, 1997). If libraries make a concerted effort to help the Web "understand" and supply the Internet with library collections, library patrons' research and scholarship productivity will be transformed completely.

PUBLISHER INTEGRATION—VALUE OF PUBLISHER-PRODUCED METADATA

Libraries and vendors should regard the Web as the destination for discovery and together improve the Collection Services Supply Chain. With that in mind, they should share the Make process and expertise. Along the knowledge making and dissemination chain, there are many separate Make processes that libraries and publishers can collaborate on. For example, publishers created a metadata standard, ONIX, to describe their publications; meanwhile, libraries created MARC21 to describe library collections. Although there have been efforts to crosswalk the fields of the two standards, the job was not always straightforward. The author argues that for better efficiency and sharing of the Make processes, suppliers and libraries should agree upon adopting one standard used along their supply chain. By doing so, libraries and publishers will not have to spend more resources on conversions between their standards version after version. For the benefit of library professionals not familiar with it, ONIX will be briefly introduced in this chapter.

According to EDItEUR, one of the principle organizations behind ONIX, ONIX is a set of international standards for representing and communicating publication information in electronic form. The ONIX family includes standards for books, serials, and licensing terms and rights information. All ONIX standards are designed to support computer-to-computer communication between parties involved in creating, distributing, licensing, or otherwise making available intellectual property in published form, whether physical or digital. All three are expressed in XML. ONIX for books was the first one and became widely adopted; ONIX for Serials provides products, subscription, online holdings, preservation holdings, release notifications, and catalog and pricing information; ONIX for licensing terms is designed for different user groups and applications, such as publishers, subscription agents, libraries, and reproduction rights organizations. ONIX for books, for example, wraps rich book metadata in XML, providing a consistent way for publishers, retailers, and their supply chain partners to share book information. As a communication format, it makes it possible to deliver rich product information into the supply chain in a standard form, to wholesalers and distributors, to larger retailers, to data aggregators, and to affiliate companies. The adoption of ONIX greatly reduces support costs, as publishers no longer need to provide data in so many unique formats. In many cases, a single data feed can be made suitable for all of a publisher's supply chain partners. And by providing a template for the content and structure of a product record, ONIX helps to stimulate the introduction of better internal information systems, capable of bringing together all the "metadata" needed for the description and promotion of new and backlist titles. In some countries, this template is used in data quality accreditation schemes. The same core ONIX data can also be used to produce advance information sheets, catalogs, and other promotional material, to feed publisher websites, and to meet the needs of the wider supply chain. As it is not designed for a specific language, it is widely used throughout the book and e-book supply chains in North America, Europe, and Australasia and is increasingly being adopted in the Asia-Pacific region.

At this point, there is little direct adoption of ONIX in the library community, although the benefits are obvious. Collins and Grogg (2011) noted that ONIX for Licensing Terms and Rights, particularly the one designed for Publication Licenses (ONIX-PL) would save libraries from laborious manual processes of updating licensing terms by automating population of general terms, but it has yet to see widespread adoption. Library closed workflows are not part of the supply chain that initiates from publishers all the way down to retailers and publishers' other supply chain partners. There is a clear opportunity to enable simple and more efficient sharing of metadata as part of the supply chain from publishers to libraries, eliminating the need for format conversions. Publishers could pass their metadata to libraries, and libraries could continue to enhance the records over time by adding additional holding and preservation metadata.

Libraries and vendors should get an agreement on one standard to share publication metadata, and thus streamline knowledge transfer from publishers to libraries, and, eventually, to end users. It will also solve many of the cataloging backlog issues that libraries are currently experiencing. Libraries are recreating much of the available metadata from the publishers in MARC21. Again, metadata interoperability between publishers and libraries should not be seen as a call to downsize Technical Services and other support functions, but rather as an opportunity to implement this shared standard initiative to free up valuable cataloging expertise to focus on describing unique and rare materials rather than general collections. Acquisitions departments will take on a new strategic role in the integration of the supply chain, as ordering will be more automated through one standard, process, and even workflow. Acquisitions should be an infrastructure enabler that drives more supply chain integration, process efficiency, and content throughputs from vendors all the way into the hands of users. Acquisitions will also function more as library representatives who are more proactively managing relationships and exchange of information between libraries and vendors. The above is a radical departure from the current job portfolios of the Acquisitions department, which remains primarily transactional in nature. The new Acquisitions department will need to improve the interaction between libraries and vendors and manage the performance of the supply chain integration.

If the standard is unified, libraries will also be able to integrate usage information using the same criteria. The author believes that there is value in distributing the preservation and usage data back to publishers as well, regardless of the nature of the collections (general, special, or archival). Such a two-way information sharing approach, one essential characteristic of a supply chain, may assist publishers with deciding on inventory and retention and help with solicitation of new book projects.

SEMANTIC TECHNOLOGY—VALUES OF SEMANTIC LINKING

Another emerging area that libraries and vendors should work together on for patron's benefit is the semantic linking of publications. Semantic Web holds a future of the Web, given its promise of replacing current document classifications and web-linking mechanisms. Library bibliographic records are flat, one-dimensional documents, which rely on subject headings and terms to create relationships between publications. Web pages are documents as well, which require a combination of techniques (metatags, frequencies of keywords, etc.) to create relevancies among Web documents.

Semantic Web is fundamentally different. It is a precise way to search and discover content as concepts are linked semantically. For example, by searching the term "windows," publications treated with semantic linking will understand whether this is the "windows" of a building or "Windows" of the Microsoft Operating System. If searching for "red," Semantic Web will aid users to navigate to the color of red or the movie *RED*. The current Web or library catalog searches will return results with a representation of many possible semantic concepts. Taking "windows" and "red" as examples, the results of search pages will not be optimized as the semantic search result pages could. Users will have to decipher the differences and sift through the results to find what they need.

Furthermore, the current search technology that relies on human interpretation produces bias and human errors. The Library of Congress Subject Headings, the standard controlled vocabulary for library materials, are repeatedly criticized for their inability to represent the finer nuances of subject relationships and the false relationships they accidentally create (Olson, 1996). The term, "male prostitutes," for example, is linked as a narrower term to "prostitutes," which in turn is linked as a narrower term for "women," creating the nonsensical suggestion that male prostitutes are women (Campbell & Fast, 2004). Semantic Web would entirely address the bias because concepts are semantically linked. More importantly, Semantic Web is a step up from human interpretation and statistical analysis (frequency of keywords, for example) to a machine-understandable structure because of the semantic treatment of the resources, which enables better machine learning and eventually arrives at better artificial intelligence.

Libraries and vendors should ally with each other to create tools that will enable employees to create semantically structured resources. This partnership should aim at transcending the current bibliographic practices into one that is fully aligned with the future of the Web. By doing so, libraries will have an incredible opportunity to change the nature of their local catalogs into a global research tool. As Web architecture keeps evolving, the author believes that it will move away from a document-centric or flat structure. Libraries, as knowledge organizations, will need to learn to grow with and adapt to the emerging standards. As an industry, libraries created very effective standards and communication protocols to share information and collections. Because of the success of the existing library network, the attitude to adopt standards that are not invented by the library community has been extremely reluctant. Integration of any external to libraries standard often ends up taking a form of conversion to current library standards. As Campbell and Fast (2004) expressed, "Metadata architectures that underlie many current digital library and Web portal initiatives are based on an assumption that information organizations, whether conventional libraries, digital libraries, or some other information body, create complex descriptions for their own specific purposes and send only simplified reductions of that complexity across the Web in metadata repositories. Such an approach is highly useful to the development of interoperability standards, as witnessed by the success of the Open Archives Initiative, which uses unqualified Dublin Core as its metadata lingua franca for cross-domain resource harvesting (Lagoze & Van de Sompel, 2001). However, although this approach goes a long way towards ensuring smooth and effective delivery of information across the Web, it does not necessarily allow libraries to exploit and contribute to the emerging Semantic Web in a full and exciting way."

Just as library standards offered the much needed consistency and stability in the past, today libraries should move beyond their own network. Because many of the library standards and practices were designed around print medium, the current standards are clunky and have a hard time evolving into the digital world.

The entire library network will need to take a new form that operates on a larger scale system by embracing newer technologies of that system. The core of the Semantic Web offers unified ways to handle library resources, whether physical or digital, through the use of Uniform Resource Identifier (URI). URI is more than a Web-addressing mechanism, as it can be employed not only for books, articles, and data set but also to concepts, such as people, places, and events, represented in library materials. The idea of namespaces and ontologies of the Semantic Web will ensure machine readability of those concepts throughout the system.

Semantic Web is important to the Collection Services Supply Chain because it presents some opportunities to revolutionize the Make processes of publications between vendors and libraries. Ultimately, the semantically enabled Make processes will enhance end users' productivity with a much better underlying search technology. The Semantic Make processes can be coordinated and shared by publishers, vendors, and libraries, passing up and down the Chain. The Semantic Make processes can be extended to incorporate others who work with Semantic metadata, which will make the process scalable even beyond the collaboration within the library community or the publisher community. Such scalability is vital for the limited human resources to address the ongoing proliferation of publications. Semantic Web is critical to the Collection Services Supply Chain because it presents some opportunities to revolutionize the Make processes of publications from publishers and libraries. It can create open identifications for anything or any concept and be open to anyone who can contribute to the creation of the semantic elements. There is enormous potential for knowledge workers from publishers and libraries to play leadership roles to ensure interoperability and to build global practices on semantic linking. The library community has been making efforts to publish their catalog records in a semantic fashion, but the utility and scalability will come with tools that support library professionals to create native semantic records. By ensuring the open and extensive linking structure, library resources and other resources can be presented together on the Web for users to get a full picture of subjects based on semantic relevancy and relationships.

REFERENCES

Butler, P. (1952). The cultural function of the library. *The Library Quarterly*, *22*(2), 79–91. http://dx.doi.org/10.1086/617872.

Campbell, D. G., & Fast, K. V. (2004). Academic libraries and the semantic web: What the future may hold for research-supporting library catalogues. *The Journal of Academic Librarianship*, *30*(5), 382–390.

Collins, M., & Grogg, J. E. (2011). Building a better ERMS: Librarians and vendors weigh in on the state of electronic resource management systems—and what still needs to be done. (LJ series: The big tools). *Library Journal*, *136*(4), 22.

EDiTEUR. (n.d.). *ONIX overview*. http://www.editeur.org/83/Overview/.

Educom Review Staff. (1997). Networked information: Finding what's out there: Clifford A. Lynch interview. *Educom Review*, *32*(6).

Hamel, G., & Prahalad, C. K. (1990). The core competency of the corporation. *Harvard Business Review*, *68*(3), 79–90.

Lagoze, C., & Van de Sompel, H. (2001). *The open archives initiative: Building a low-barrier interoperability framework*.

Olson, H. (1996). Between control and chaos: An ethical perspective on authority control. In *Authority control in the 21st century: An invitational conference*.

CHAPTER 15

Start With the End in Mind

KEEPING USERS SATISFIED IS PRIMARY FOCUS

Throughout earlier chapters of the book, the author repeatedly proposed that supply chain should always keep end users in mind. When starting to design or implement a supply chain, libraries not only should have their patrons in mind but more so they should begin by engaging and involving and partnering with their patrons. When process solutions are recommended, departments along the supply chain should provide feedback and test those solutions. Vendors should be involved in the process as well, which was elaborated on in Chapter 14. Eventually, libraries must recalibrate their current processes to meet user needs by aligning vendor processes to those of their internal departments.

In earlier chapters, the author also mentioned recasting the relationship between vendors and libraries, libraries and users as partners rather than simply seeing their relationship as transactional, buy–sell, and supportive-type only.

It is easy for libraries to see processes and workflows as their end goal when coming to supply chain improvement because the term "supply chain" literally gives too much emphasis on processes. Libraries should avoid any process improvement without considering the context of the entire system and end results that benefit user teaching, learning, and research productivities. Chapters 16 and 17 will discuss the system view of the supply chain improvement and provide some tactical suggestions for libraries to implement the supply chain concept. This chapter will concentrate on outlining practical ways of starting with user needs in mind for supply chain design and change.

Do libraries know what their patrons desire most from library collection services? The simple answer is to ask them, which libraries absolutely need to do. Throughout his professional career, the author has been promoting customer-centered techniques to improve library services. In one of his previous positions, the author managed a Web services team. That experience provided deeper understanding of reasons to focus improvements and implementation of new services around patron needs. In the world of the

Supply Chain Management for Collection Services of Academic Libraries
ISBN 978-0-08-102031-9
http://dx.doi.org/10.1016/B978-0-08-102031-9.00015-3

Web, notion of user experience is essential. The Web community has developed many approaches to studying user preferences and behaviors. Such tools and techniques, as personnas, ethnographic studies, surveys, Web analytics tools, usability testing, and focus groups, are used to learn about user behaviors and preferences. There is no one single tool that can provide a holistic view of users, but a synthesis of findings extracted from multiple sources suggests a near-reality picture of users.

Libraries have been practicing the same techniques listed above for many other services with the exception of collection services. Some libraries have been leveraging circulation and interlibrary loan (ILL) data to inform collection development, analogous to using Web analytics to understand user click-throughs and navigations; however, not many other techniques were employed in the area of collection development. Selectors seldom engage academic departments they liaise on collection topics, so the collection development activity defaults to buying and building a body of knowledge. No one questions why libraries keep buying more and more and how that body of knowledge translates into being relevant to user research and scholarship. When starting with users in mind, libraries need to understand their patrons' research and scholarly productivity. How do library collections contribute or not to their productivity? How can libraries do better with the resources they have?

Even with the assistance of a variety of user studies, the author found that analysis of and any improvement to user Web experience could be a guessing game. Sometimes, even after the requested improvements are made, they still do not address original user concerns. Occasionally, the author walks away from conversations with users very puzzled questioning whether they know what they need. This varied personal experience informed the author's beliefs that despite all the user studies techniques, libraries should be looking inwardly and question the fundamental goal of user research and learning activities. For collection services, in the author's opinion, what users are expecting from their libraries is not the size of the collection, but a recipe for perfect collection fulfillment. A Perfect Collection Fulfillment is defined as follows:

1. Identify the needed content
2. Review all possibilities (complete body of literature)
3. Pick a preferred format (digital or analog)
4. Interact via a preferred interface (phone, Web, or in person),
5. Select a preferred retrieval option (request for delivery and pickup, including ILL, or self-retrieval at the stacks)

6. Obtain the right content (language, edition, etc.)
7. Deliver to the right destination (office, home, device, pickup location, and stack)
8. Deliver at the expected time (delivery service agreement, on stack not missing)
9. Deliver in the right condition (usable)

The absence of any condition from the list above results in an imperfect collection fulfillment, causing dissatisfaction and frustration among users, creating inconveniences to users, and impacting user productivity. An imperfect fulfillment can be costly to libraries, not only in reputation but also in collection budget, staff time, and other resources. Libraries should sincerely ask themselves whether the Perfect Collection Fulfillment is the collection experience that they desire to offer to their patrons. The author is not sure that some patrons would ask libraries to do precisely as described above, but are libraries willing to make a commitment to their patrons to build a Perfect Collection Fulfillment process? Starting with the end in mind means the end is the perfect collection fulfillment experience.

The perfect fulfillment consists of two parts: discovery experience and use experience. First of all, patrons have to be able to find what they need. At the discovery stage, libraries only need to provide access to information about existing publications, rather than acquire those publications. To forecast is difficult, which is the reason why Part two—Library Supply Chain Model of the book suggests libraries segment their collections into Must-have, Good-to-have, and specialized collections. The book also recommends prioritizing Make-to-User model. The author believes that at this point of discovery, libraries only need to surface the publication information without acquiring most of the actual items. Libraries should provide as much publication information as possible, and by doing so, libraries will meet Points 1, 2, and 3 of the Perfect Collection Fulfillment. The above is the reason why the author recommends using shared metadata standard and workflow between publishers and libraries. Shared process is an economical, sustainable, and scalable practice to provide rich publication metadata, such as description, holding, formats, and pricing information. Because of the aggregation of all possible publication data, patrons will have confidence of adequate coverage in their field and thus pick the right content for their needs.

Points 4 and 5 offer the experience that users prefer to have. Even in the Digital Age, libraries should respect their users' preferences. Often, libraries push acquisitions to preferred electronic formats because of space and budget constraints. The right collection decision should be made in accordance with user goals and needs. Points 6–9 concern the use experience. Collections should be found on shelves where the catalog says they should be or be delivered to the location of the user choice. What users find on the shelf or obtain via delivery should be an exact match of what users requested and in a condition that is usable. Finally, users should get what they need in a timely fashion. Often, users go to the stacks and only find the desired items to be missing or misplaced. Only if libraries meet all nine requirements of the Perfect Collection Fulfillment will collections reach the level of service excellence.

WORKFLOW AND PROCESS IMPROVEMENT GOALS ARE SECONDARY CORE

With the Perfect Collection Fulfillment as the ultimate goal, libraries will be able to measure the performance of their Collection Services Supply Chain, tracking backward from Return, Delivery, Make, Source to Plan processes. Libraries can analyze how often they perform perfect fulfillments. The reverse thinking can provide insights into identifying root causes of falling short of perfect fulfillment and guide libraries toward lessening or eliminating service issues. The author recommended some metrics in Part two—Library Supply Chain Model of the book, many of which are Key Performance Indicators (KPIs) to identify potential collection process issues along the Supply Chain.

Although KPIs improve task efficiency and productivity, libraries must keep in mind that the processes serve the ultimate end goal: user productivity. Process improvements should not be undertaken just for the sake of making changes, particularly if those improvements are performed strictly in the context of a unit and a department. Libraries should work on developing profound and broad relationships with scholars and researchers to understand their collection needs first. Such long-term depth and breadth of a relationship will help libraries collect the needed critical knowledge from their patrons. This relationship can continue to guide process improvement efforts. Sometimes, users may offer information that will sequentially change a process in its entirety.

Current collection development processes remain quite selector-centric. Selectors' personalities, social networks, bias, interests, professional training, and their ideologies significantly influence those processes. Although selectors acquire collections based on general university teaching and research agendas, their engagement with patrons of the broader academy is insufficient.

Lee (2003) conducted research that examined and analyzed the collection development processes based on institutional and social developments in the context of the library science discipline. Lee's study exposed a need for research to better understand the advisory role played by library collections. In his opinion, without such a research many critical collection development questions could not be answered. For example, could extensive and comprehensive research collection for all disciplines be more effective in serving users in this interdisciplinary field? Lee also pointed out the limitations in understanding collection demands by analyzing search behaviors associated with specific tools and resources and deduced from circulation and statistics data. According to Lee's research, it is unclear how a collection of information resources in and of itself forms a context to influence users' information-seeking behavior.

To be fair, the problem does not reside solely within libraries, as academic departments often hold a biased perception of collection services. Not rarely, collection services are viewed by academic departments as being about building a large collection body. Such opinions are typically formed without relying on basic understanding of any of the underlying economic and operational details of library business. So a meaningful relationship between libraries and the rest of the academy is even more essential going forward. As collection stewards, libraries will need to make their collection practices, statistics, and budget information as transparent as possible. Only through such level of exchange of information, libraries and academic departments can reach a wider strategic direction for collection development and move toward demand-centric collection supply processes. The demand-centric methodology should benefit all departments, not just a few. The critical information gathered from research on academic collection requirements must be analyzed by the right functional area in libraries and more importantly passed on to the appropriate people to follow through. Chapters 16 and 17 will propose ideas on how to break siloed approaches and how to implement changes to supply chain processes.

FINANCIAL AND RESOURCE STEWARDSHIPS ARE BOTTOM LINE

Although the author advocated that libraries should act more like businesses, he has no intention or desire to prioritize budgetary goals, assets and liability balance sheets, or financial numbers as the goals of libraries. Libraries are cultural and archival institutions. For academic and research libraries especially, the main objective of the business should be meeting needs of their diverse user groups. Those user groups can be categorized by disciplines, research interests, identities, and a variety of other methods. One of the author's mentors once coached him by saying that "libraries have to be everything to everybody." Although financial gains should never be on the list of the priorities, balancing the budget will always remain a critical task for libraries. The author sees the industry as being "hijacked" by the job of keeping ends meet involuntarily, and as a result, losing its focus on the purpose of resource stewardship. Many cancellation of subscriptions or shift from print to e-preferred strategies were not motivated by faculty requests, but were instead imposed on libraries by pressure from price increases and storage constraints. It is not that cancellations and e-preferred strategies were wrong moves. In the author's opinion, they should have begun much earlier than they did, but not as a reaction to immediate crises, but instead as a proactive planning strategy based on research of collection needs.

If libraries have to be everything to everybody with their limited resources, careful and strategic resource planning is a must. Currently, planning is done based on established prior commitments, so it is unlikely that libraries can meet emerging needs. For some institutions, with ongoing budget cuts, even support of current needs has become curtailed, and therefore, to support new requirements is a very daunting task. The current collection planning model pushes libraries into a vicious cycle that consists of trying to sustain commitment to current needs, often struggling to keep up with them, and as a result, having no ability to allocate resources to address additional new needs.

To serve everybody should not translate into a suggestion that each user group receives the same share of the pie. Throughout his career, the author worked in institutions where each library department was subject to equal percentage of funding cut in response to budget reductions. It was "fair" from the point of sharing mutual burdens, but it stalled several strategic growth areas, especially those where resources were initially thin. Budget accountability comes from well-planned collection needs and consequently from well-executed acquisitions, collection description, conservation, preservation, delivery, and return processes. Libraries have to move away from the practice of "guessing" or making assumptions about what their patrons need. Libraries should start with a master "blueprint" of collection services to avoid making collection decisions guided by either prioritizing financial bottom lines or other priorities of the time.

At the same time, different user groups might provide libraries with various contradictory ideas about their desired collection services. For this reason, the author will introduce two popular methods that will help libraries create a master "blueprint" for serving all users. The first one, Quality Function Deployment (QFD), is a well-adopted technique for product or service design. QFD may sound foreign to libraries, but many perhaps have used pieces of it to surface patrons' needs. Customer Voice is one of the QFD methods that libraries have adopted. QFD is a technique that is used to translate requirements—defined by customers—into specifications for a product or service (Ayers, 2001). The quality of collection services must be thoughtfully designed rather than piecemealed together. A successful collection service is not only represented by the content itself, but more importantly, by many of the supply chain features. The collection service excellence composes of many ingredients from supply chain Plan, Source, Make, Deliver, to Return processes. What do libraries supply to their patrons? How quickly do libraries process those collections? How well do libraries describe, preserve, and curate those collections? How well do search tools function? And last, but not least, how quickly do libraries locate requested collections on the shelf or how fast do libraries deliver them? What libraries may learn from QFD has profound implications for management processes of the Collection Services Supply Chain from design of the Chain to its implementation. Part two—Library Supply Chain Model of the book introduced the concept of the "extended product or services." Supply Chain features are a critical part of the "extend" part of the product or the service. Advantages of QFD reside in its synthesis of a variety of channels used to collect user feedback, such as surveys, focus groups, statistics, and studies; methods that help translate "what" into "how." Often "what" is uniformly well understood by libraries, but "how" is the area where differences arise. For example, a faculty wants to keep up with the new development in his or her field. How will libraries understand the need and translate "what" into "how"? Does it mean libraries have to subscribe to and buy a large body of collections in that discipline? Or does it mean libraries should aggregate publication metadata so that the faculty in question can conveniently search against it? How frequently does he or she do it? How quickly is it needed? Currently, many libraries' practice has been to buy and license content. Libraries seldom slow down and ask why they buy what they buy. The synthesis of gathered user needs data forces libraries to confront the bias or assumptions inherited from those feedback channels and to have honest conversations to understand user needs, prioritizing patron requirements for further planning and implementation. QFD is quite likely to raise questions about libraries' current supply chain processes, which may be found in conflict with the design outcomes of perfect collection services. If libraries decide that the Perfect Collection Fulfillment is the end goal, there could be many processes that would need to be redesigned. Because there are many existing works of literature on QFD and its tools and implementations, this chapter will refrain from providing further details about this method.

The second approach that libraries may consider is called Qualitative Heuristic (QH). The method, developed at the University of Hamburg, Germany, attempted to bring back the qualities of systematic exploration and discovery into psychological and sociological research (Kleining & Witt, 2000). The quality aspect of systematic exploration was later applied to information systems and software engineering, where the author started his interaction with this methodology. It made much more sense to the author because QH helped him understand the psychological and sociological components of systems and software design later; although initially, it was a complicated method to interact with intellectually, and it proved to be mentally demanding. Because QH calls for deep introspection, it requires IT professionals to take extra steps toward understanding and interpreting subjective qualities, which could be a very counterintuitive burden in addition to the regular systems and software development tasks. Often, QH causes conflicts between ideal IT development, processes, and features with that of subjective qualities. Maybe a better way to put it is the subjectiveness of the IT professional's education or training conflicts with the subjectiveness of user-desired qualities. One good example applicable to the library industry is FRBR (Functional Requirements for Bibliographic Records), which is an exceptional way to represent construction and relationships of bibliographic and authority records. However, speaking from the author's experience of its real-world implementation, FRBR often causes issues for users who search and try to locate the right item in discovery tools. FRBR is a useful framework for the library profession, but because patrons' search behaviors are rather subjective, patrons' goals come in conflict with the design principles and implementation of FRBR. These types of discords between customer goals and desire for professional perfectionism are not uncommon. Professionals should not see them as right or wrong issues, but rather as issues of what is working and what is not working for users. QH pushes libraries to confront our established and deeply rooted professional practices, ideals, and assumptions against those of our users, which makes QH an extremely valuable approach for designing collection services and its supply chain processes. Kleining and Witt (2000) outlined four ground rules for QH:

1. The research person should be open to new concepts and change his/her preconceptions if the data are not in agreement with them.
2. The topic of research is preliminary and may change during the research process.
3. Data should be collected under the paradigm of maximum structural variation of perspectives.
4. The analysis is directed toward discovery of similarities. It locates similarities, accordance, analogies, or homologies within these most diverse and varied data.

Of course, libraries may apply other methodologies or techniques listed in this chapter, to practice the principle of "start with the end in mind," or "start with the user collection needs first." As long as libraries understand that the goal of the Collection Services Supply Chain is not a strategy for maximizing process efficiencies and lowering costs, there are many other paths that will help libraries align what they do with what their users need.

REFERENCES

Ayers, J. (2001). *Handbook of supply chain management.* Boca Raton, FL: Alexandria, VA: The St. Lucie Press/APICS Series on Resource Management.

IFLA Study Group on the Functional Requirements for Bibliographic Records. (1998). *Functional requirements for bibliographic records: Final report.*

Kleining, G., & Witt, H. (2000). The qualitative heuristic approach: A methodology for discovery in psychology and the social sciences. Rediscovering the method of introspection as an example. *Forum: Qualitative Social Research, 1*(1).

Lee, H. (2003). Collection development as a social process. *Journal of Academic Librarianship, 29*(1), 23–31.

CHAPTER 16

Whole Is Greater Than Sum of Its Parts

SILO IS THE NORM

Functional hierarchical line management has been the dominant management paradigm in management theory and practice according to Turner and Keegan (1999). Not just libraries, many organizations were formed by functional groups and the respective subfunctional groups. According to the organization principles of Fayol, Urwick, and Taylor, the functions of an organization ideally should be clearly bounded from one another (Buijs & Edelenbos, 2012). Line managers of an organization are accountable for integrating and coordinating its domains and sectors (Kanter, 1983). However, one major issue of this management paradigm is its ineffectiveness to integrate and coordinate across the functional lines (Koteen, 1997). To compensate for this shortcoming of the management model, many have introduced matrix approaches, establishing committees, working groups, task forces, and teams to blur the functional lines and tackle systemwide issues. Libraries were no exception to introducing matrix elements to their line management, coping with siloed operations and boosted functional integration. The results vary. One of the author's friends once said that it was easier to collaborate across campus and even beyond the institutional walls rather than within departments in libraries. (This statement is probably true for many hierarchical organizations.) Libraries are largely fragmented: departments behave like mini-kingdoms of a federated empire, exercising self-serving autonomy and freedom. Line managers often do not exhibit an appreciation of librarywide strategic endeavors within the scope of their own operations.

This phenomenon is very easy to understand from the practices angle of the hierarchical line management. Budget is allocated to each department, and positions are designed to carry regular work of a department. The above practice isolates departments from one another, reducing interdependencies and needs for communication and collaboration. As departments have full control of their operations, they develop best departmental workflows, processes, procedures, and even agendas, which are not necessarily in line with

Supply Chain Management for Collection Services of Academic Libraries
ISBN 978-0-08-102031-9
http://dx.doi.org/10.1016/B978-0-08-102031-9.00016-5

others or the organization as a whole. Without proper oversight, departments quickly develop tunnel vision, tuning out concerns of others and those of the entire system. As lean as library resources are, library administrators often have to take away resources from one department to supplement the needs of another. As many of those transfers of resources are strategic, from line managers' point of view, it creates great levels of stress for units that lose resources and makes it challenging to sustain their routine operations. When line managers request new and additional positions for their teams, many, lacking the sense of an overall growth and strategic direction of their organization, often feel they are in competition rather than in collaboration with other units. It is as if they are chess opponents rather than teammates who are on the same side of the board trying to win the game together.

Supply Chain Management (SCM) approach will not be able to change this siloed environment. It requires senior library management and functional managers to transcend organizational line structures and work in close cooperation together. Without intentional coordination across all departments, supply chains remain a theory on paper. At least initially, change must begin with library leaders creating a collaborative atmosphere for the adoption of SCM. Once it gets a jump-start and as employees and users begin to see benefits, SCM will take over driving collaboration further and deeper in libraries.

The author is not suggesting libraries should abandon their management structure and completely switch to a different model. It is up to the leadership of an organization to drive the integration and empower employees to work collaboratively, as other management models also have inherited shortcomings as well. For example, it is hard to reach a leadership consensus in an open and flat governance model. One thing that libraries must change though to promote empowerment is to eliminate bureaucratic controls as much as possible, so decision making takes place at the right level and at the right time.

ISSUES OF SILO OPERATIONS

Blankenship and Miles (1968) conducted a very interesting study on organizational structure and managerial decision behavior. In general, manager's location in the hierarchy appeared to be the major determinant of behavior patterns. Managers at the top of the hierarchy usually enjoyed the most decision autonomy, while at the same time, they also relied heavily on subordinates to bring problems to their attention and to offer ideas and recommendations for solutions.

For lower level managers, the pattern of decision behavior was the opposite. They did not rely on their subordinates for initiation of action or submission of suggestions, but took their guidance from above. In addition, less than one-third of them felt that their recommendations to their superiors carried much weight; yet at the same time, it was rather surprising to find that 41% still felt that they enjoyed having the right-of-final choice without prior approval from above.

Although it was a dated study, it probably still reflects some truth for today's libraries. It poses an interesting challenge, doesn't it? If issues are not reported up, the upper management would rarely have a chance to know about them and even less probability to act on them. As some line managers enjoy their decision-making autonomy, there could be little to no motivation to surface problems. After all, reporting issues up will force them to give up their decision rights. So how should libraries address these behavior issues? Whatever the solution might be, it should not force libraries to become more bureaucratic by centralizing all decision-making power at the top. Top managers do not have to initiate most of the decisions; instead, they should work on increasing the overall institutional competence level in decision making. Libraries should acknowledge these behavior patterns and leverage the decision autonomy for good decision making. Libraries should consider creating proper mechanisms to surface recommendations within and outside their former structure and to grow the ability of employees to make the right decisions at their appropriate levels.

When discussing improvement of institutional competence in decision making, Borgio (2003) pointed out disadvantages of a rigid hierarchical organizational structure and recommended an alternative arrangement based on process-oriented functional decision making. The process-based method offers an opportunity for departments to communicate based on processes across the lines of their operations. Also because process is a fundamental component of SCM, this style of decision making provides guidance for conversations on deficiencies around departmental handoff issues around processes and integrates their processes into a cohesive and synergistic one that benefits patrons.

INTEGRATION IS KEY

It is important to note that SCM is a cross-departmental discipline, which needs participation from both horizontal and vertical levels in libraries. Collaboration throughout the libraries is a prerequisite for producing meaningful changes to collection processes and eventually to collection services. The best way to implement the Collection Service Supply Chain is to prioritize responsibilities for managers by including management of supply chain integration across the institution in their responsibilities rather than only asking them to focus on their own groups' processes. The previous statement does not mean to assume that frontline employees are not critical for supply chain integration; yet, managers perform a vital role and have the authority to initiate many process changes. Managers should be the first group to be brought on board, which could be challenging; however, if it is executed well, libraries have already overcome one of the biggest possible barriers and would gain momentum to forward the supply chain implementation. Getting managers on board is a sure way to realize the strategic significance at the daily operational level, which, in the author's opinion, is the only way to achieve substantial and concrete results.

As with introducing any new concept to the workplace, library administrators need to approach this from change management perspective and create steps to promote and implement changes. This is true especially for supply chain implementation because most libraries will have first-time experience with the concept. There are no references and previous experience that libraries may capitalize on and borrow from.

When the SCOR model was first introduced in Chapters 4 and 5, the author deliberately left one of the processes, Enable, out from the entire Part two—Library Supply Chain Model of the book. The reason for that decision was not because Enable does not fit within the library environment, but because Enable processes can be the guiding force behind informing decisions for any supply chain changes or implementation in libraries. Supply Chain Council added it into the most recent update of the SCOR model in version 11. The author saved it until now to introduce it in this part of book, which deals with institutional culture changes and supply chain innovation.

SCOR has been used for a long time as a standard approach to design and implementation of supply chain practices and business processes by industries and some of the well-recognized global brands. The update of SCOR 11 is a significant milestone, as the previous versions included nested process steps, guiding institutions to design their processes and metrics. According to Newman's article on the techtarget.com, in SCOR 11, the Supply Chain Council review team, which comprises supply chain practitioners who walk the walk and talk the talk, noticed that a common Level 2 process called Enable was worthy of being moved up to Level 1 status. The Enable Level 1 process provides guidance on how to assist the initial five process steps: Plan, Source, Make, Deliver, and Return, with best practices to support those processes. This change essentially creates a closed-loop model to SCOR 11, similar to the current ISO 9000 standards built on the Deming "Plan-Do-Check-Act" feedback model, which will be introduced later in the chapter as a tool that manages change.

First, SCC (2010) defines Enable as "The process associated with establishing, maintaining and monitoring information, relationships, resources, assets, business rules, compliance and contracts required to operate the supply chain. Enable processes support the realization and governance of the planning and execution processes of supply chains. Enable processes interact with processes in other domains (for example financial processes, HR processes, IT processes, facilities management processes, product and portfolio management processes, product and process design processes and sales and support processes)." Enable provides all the aspects that an organization should consider implementing to facilitate monitoring the progress of a supply chain. It very much focuses on the integration of those areas of the supply chain and the integration with other critical processes that an organization creates to support its operations.

Then comes the "Plan-Do-Check-Act" feedback model. Based on Ayers (2001), Deming taught the Shewhart cycle of Plan-Do-Check-Act in Japan. The Shewhart model provides a continuous, durable, and effective way to upgrade a process. For libraries, it is important to start the supply chain with continuous improvement rather than to strive to set up a comprehensive framework right from the first. In author's opinion, libraries have gone through so many "waterfall" approaches to creating a perfect solution only to see either the environment shift on them or fall into the analysis-to-paralysis state. Note that the Shewhart Model is a closed loop, so the feedback in practice always goes back to the Plan stage for recalibrating the next round of Plan-Do-Check-Act.

In the Plan phase, Ayers (2001) suggested identifying connections of the supply chain to overall strategies of the organization, constituting the concept at a very high level. A portfolio of supply chain–related initiatives and projects should also be created. In the Do phase, it is advisable to write project action plans, test solutions in pilot environment, and implement the organizational structure and measurements needed for full-scale implementation. It is also necessary to use appropriate technology to support the redesigned processes. In Check, the suggestion is to observe the results and change the solutions accordingly. In Act, organizations should evaluate the change, learn from the results, and begin the next planning cycle.

INTEGRATION AT THE TOP

The Shewhart four-step process is not an unfamiliar model within the library industry. Many have adopted the idea of pilot testing of new initiatives. Shewhart Model can push pilots further in libraries because it provides an iterative approach and offers techniques to deal with unsuccessful results. Some pilots of initiatives most likely failed in the past not because they are not valuable to patrons or to the institution, but because institutional culture failed them. Shewhart Model particularly emphasizes the importance for organizations to adapt their structure and measurements to implement a full-scale implementation, which requires top management to play very specific roles. Integration at the top management level is absolutely essential. Going back to the study by Blankenship and Miles, senior managers seldom initiate work of their own and in most cases rely on subordinates to come up with recommendations. In the case of Supply Chain implementation, senior management has to launch the initiative and create Enable process to guide the organization, managers, and employees through the change.

Top management's buy-in, full support, and their involvement as sponsors of supply chain initiatives are rigid preconditions for success. As sponsors, senior management should not make all the decisions, but they should create an environment that empowers and compels employees to make right decisions at every level along the way. Sponsors should attend to infrastructural issues that either enable or cripple change. Sponsors are ultimate owners of supply chains, so they need to create clear vision and expectations, advocate for resources, set up priorities, and create strategies to arrive at the results of their supply chains. A sponsor has to be aware of any bumps along the way, not personally solving them, per se, but understanding them, guiding the organization to adapt and calibrate along the way. Without the senior level of engagement, implementers will get lost and worn-out quickly. Senior management has to eliminate structural bottlenecks in organizations, create an empowering and learning environment to maximize the most potential of every employee, and set up measures and reward systems that change behaviors. Integration at the top administrator level is extremely important as the management layer controls access to resources and owns the ultimate responsibilities to make libraries successful.

INTEGRATION WITH THE REST

Once top-level management fully engages in the supply chain concept, the second tier of integration will be with library departments. The main obstacles for implementing supply chain at this layer of integration are learning and applying the concept, framework, and tools. Even if departments see value of the supply chain approach, a closer-to-reality scenario would be that they experience a clash of preferences, processes, procedures, and tasks that occur between supply chain concept and their own. Top management should respect the fact that libraries have been structured into separate functions, a norm based on which libraries have been operating. Senior leaders should also recognize the types of "functional expertise" that have been built up within each department. For libraries, those functions or departments can be represented by acquisitions, cataloging, circulation, reference, and subjects. Although the names of the above functions were rather conventional, today's libraries more or less continue being organized in a similar fashion. With time, those functional names have been changing depending on the predominant assignments of the functions. For example, reference service these days may be called public services or instructional services. Reference's primary focus has been switching from stationary or terminal services to more proactive outreach services with a goal of teaching patrons to use library resources. Another example is subject librarians. Today they are often referred to as liaisons, given their primary assignment to manage relationships with their respective academic departments. Nevertheless, the core goal of those jobs has not been evolving as library administrators would have hoped. There has been a high level of inertia and desire to cling to the old way of doing things. For example, liaisons keep buying collections rather than actively building relationships; catalogers continue cataloging with machine-readable cataloging rather than learning others metadata schemas, etc. Again, the study of Blankenship and Miles (1968) demonstrated very disconnected management layers in hierarchical organizations not suitable for change making, where top management relies on subordinates for recommendations; middle management does not include their subordinates in decision-making process and very much prefer their decision autonomy to remain intact. The author has been very fortunate in having opportunities to work as part of different functional departments in libraries, both within public services and technical departments, but many of his peers have never moved from one department to another in their entire career, thus limiting their breadth of view and familiarity with challenges that are present in various parts of a library. As a matter of fact, many professionals dive straight into a position without being provided any understanding of how their tasks relate to the "larger picture." Over time, their knowledge and expertise in a specific area grow, but their appreciation for systematic connection of their work to the larger organization tends to diminish. To become successful, the supply chain must be a librarywide endeavor and a larger organizational strategy than a functional one. Top management needs to be aware of these long-lasting patterns and find ways to work with them when looking at supply chain adoption and implementation. Top management must direct and motivate all internal functions toward participation along each step of the implementation.

The best way to move forward with department integration, in the author's opinion, would be to have a neutral party to assemble a team of functional experts from supply chain departments. The neutral party can be a project manager at the library or an outside consulting firm. It would present a problem if a functional manager were to be designated as the supply chain leader. Unless the functional manager holds a much broader view beyond his or her department, is granted the ultimate authority, and has the full support from the top administrators, it would be unlikely that a supply chain project will gain any traction and make any progress. Another proposed strategy that libraries should implement is to push the adoption of more customer-centered service philosophy, tactics, and tools in their workforce, which was explored as a separate topic in Chapter 15.

The newly assembled team would require some unique qualifications to work with the rest of the organization. In the author's opinion, team members have to be functional experts. In-depth understanding of ins and outs of their departmental processes is a mandatory requirement for anyone looking for ways to improve them. Those functional experts can intelligently speak about their processes and provide adequate information for the team to find the best solutions. Team members must be open minded, analytical, action driven, and result oriented, as such characteristics are bound to push the team toward making progress even in cases when the team encounters ideological, preferential, procedural, and methodological roadblocks. Finally, team members have to be competent communicators because they need to bring recommendations back to their representing departments for feedback and eventually be ready to champion implementation.

REFERENCES

Ayers, J. (2001). *Handbook of supply chain management*. Boca Raton, FL; Alexandria, VA: The St. Lucie Press/APICS Series on Resource Management.

Blankenship, L., & Miles, R. (1968). Organizational structure and managerial decision behavior. *Administrative Science Quarterly*, *13*(1), 106–120. http://dx.doi.org/10.2307/2391263.

Borgio, G. (2003). PMI: Process oriented hierarchical/functional organization. *Macchine Utensili*, *24*(5), 114–126.

Buijs, J., & Edelenbos, J. (2012). Connective capacity of program management: Responding to fragmented project management. *Public Administration Quarterly*, *36*(1), 3–41 Retrieved from:http://www.jstor.org/stable/23209575.

Kanter, R. (1983). *The change masters: Innovation for productivity in the American corporation.* New York: Simon and Schuster.

Koteen, J. (1997). *Strategic management in public and nonprofit organizations: Managing public concerns in an era of limit.* Westport: Praeger.

Newman, W. *SCOR 11 goes closed-loop with new release.* Retrieved from: http://searchmanufacturingerp.techtarget.com/feature/SCOR-11-goes-closed-loop-with-new-release.

SCC. (2010). *Supply chain operations reference (SCOR) model overview – version 10.0.* Cypress, TX: Supply Chain Council, Inc.

Turner, J. R., & Keegan, A. (1999). The versatile project-based organization: Governance and operational control. *European Management Journal*, *17*(3), 296–309.

CHAPTER 17

Where the Rubber Meets the Road

This book has been designing and promoting ideas of a new Supply Chain model for collection services, so it is only fitting that the final chapter of the book outlines a potential road map for those libraries that decide to adopt Supply Chain at their institutions. The path that the author proposes has much to do with personnel development and culture change, precisely the set of values that motivate people to pursue service excellence. An appropriate mind-set shift is a must because it is going to depend on individuals making a decision to play a role in either leading or contributing to the effort. The study by Fawcett, Magnan, and McCarter (2008) revealed that all managers recognize technology, information, and measurement systems as major barriers to successful Supply Chain collaboration with the people issues—such as culture, trust, aversion to change, and willingness to collaborate, being most intractable of all.

Part two—Library Supply Chain Model of the book discussed "clashes" of values between Supply Chain and Collection Development and Acquisitions. The author offered a framework for libraries to adapt and customize the Collection Services Supply Chain based on their unique environments because Supply Chain Management is not a theoretical study, but rather a system of day-to-day practices and continuous improvement projects. The only way to make Supply Chain become successful is through regular daily alignment at the operational level and finding the right people who understand the details of their tasks and who remain attuned to the purposes of all tasks along the Supply Chain continuum so incremental improvements could be implemented.

PROBLEM SOLVING VERSUS TASK-SPECIFIC WORK

Library work can be highly tasked oriented. The issue is universal for many long-standing industries—workflows and processes are well defined, and positions are designed to take on specific tasks. Because the advancement of technologies has shaken the stability of well-established industries, many of the rigidly designed duties are becoming antiquated faster than ever. At the same time, employees who have remained in fixed roles for too long are finding themselves struggling to adapt to the changes that are happening

Supply Chain Management for Collection Services of Academic Libraries
ISBN 978-0-08-102031-9
http://dx.doi.org/10.1016/B978-0-08-102031-9.00017-7

around them and to perform new sets of tasks their jobs require. People rarely take initiative to consider the driving forces behind changes to their formerly stable unchanged assignments and seldom ask why those tasks are being done and what should be changed. The task-based position philosophy greatly impedes on institutions' ability to advance and to adapt to contextual shifts. Libraries will need to transform their employees from being strictly task operators into becoming analytical problem solvers.

To successfully introduce Supply Chain and to plan for its implementation, in particular, libraries need to recruit problem solvers from within their current workforce and turn them into champions of the Supply Chain idea. More importantly, organizations should call upon those individuals to help the rest of the employee body accept the concept and prepare for its adoption. Roya Rad (2013) listed 10 characteristics of good problem solvers, which might be helpful for libraries as they work on identifying the right people for the job. Typically, problem solvers can identify root causes of symptoms and thus articulate and define problems. They can develop a plan containing clear objectives and outcomes to solve underlying causes of the issues. They are open-minded, exploratory, flexible, creative, effective, and resourceful people. They are able to work through rising obstacles and adjust their plans accordingly by finding alternatives to achieve the set objectives. Problem solvers are mindful and intentional about communicating their detailed action plans to various constituencies. They understand who stakeholders are and their associated decision rights and power. They can get the right information to the right group of stakeholders at the right time and therefore form the necessary consensus to move forward. Problem solvers always keep quality, consistency, and cost in mind, as all three define the outcome of any Supply Chain project. Problem-solving skills not only are exceedingly critical for the benefit of growing healthy internal relationships but are also vital when engaging patrons and vendors across the Extended Supply Chain. Most importantly, problem solvers whom libraries recruit for their Supply Chain idea implementation must possess thorough knowledge of library operations. They will be tasked with examining existing collection processes and required to find ways to optimize them. Ideally, they would also be familiar with the practice and knowledge of Lean Management, also known as Lean.

WASTE ELIMINATION VERSUS COST REDUCTION

Supply Chain is not a strategy for creating cost reduction, but should be employed as a means of increasing service quality through waste elimination. Senior management in libraries must first recognize and accept the significance of Supply Chain to their organizational mission and service quality. Supply Chain elevates roles fulfilled by internal functions to the same level of importance as public-facing services. Often, libraries underestimate the importance of their supporting units and view them as units associated with cost to the organization in many cases. Frequently, it is the internally facing departments that are subject to resource cuts. Either vacant positions become repurposed elsewhere in the organization or are eliminated completely, yet decisions to do so are often rooted in lack of knowledge of capacity in affected units and how those cuts impact their roles in overall Supply Chain. From the service quality point of view, all library functions, either internal or external, are essential to the ultimate goal of creating and sustaining patron satisfaction. Internal departments have the same impact on user services as public-facing units do. Cost reduction should never be the goal of any Supply Chain effort. Libraries must zero in on inefficiencies in their existing processes across all their functions and aim to streamline those processes across overall Supply Chain.

Eliminating waste and its associated cost are part of the Lean philosophy. Waste can be quantified by time, labor, or money. The concept of Lean originated in manufacturing world. Lean additionally benefits libraries and could be expressed in reduction of collection inventory and increase of resource flexibility. "Lean drives continuous improvement, which aims at improvement in small, incremental, continuous steps that can be placed in the context of a number of tactical initiatives. Daily improvement in small amounts carried out in every job and function of the business eventually accumulates into very large gains (Burton & Boeder, 2003)." The continuous improvement in small-steps approach fits within library culture very well. To implement such "radical" ideas as Supply Chain in libraries, the author recommends that organizations consider well-designed and well-executed incremental small wins that accumulate and eventually lead to planned and envisioned transformative successes.

Libraries should carefully identify and work on resolving critical bottlenecks in their collection processes. As libraries eliminate waste in their operations and people begin to enjoy regained financial resources and productivity, libraries will achieve momentum and buy-ins to move their organizations toward change further and quicker. According to Keller (2008), "Some people seem to think that continuous improvement is about finding that silver bullet and they engage in multiple kaikaku (radical change) events looking for a quantum leap in performance. Meanwhile, some of the most respected companies in lean manufacturing continue on their steady course of improvement using lots of mini kaizens performed continuously by small teams of value-adding operators right at their worksites." Keller continued, "while it's nice to have some quantum leap improvements, it is the steady continuous improvements that will yield lasting performance increases. It is also the only way to get lasting culture change in the organization and to get everyone's buy-in on 'that's how we do things around here'."

Toyota is credited with invention of the Lean Manufacturing concept. Toyota model was so successful that many other automakers and industries adopted it in their organizations with some variations. Toyota is a good case for libraries to examine, as they apply Lean across every process of its plants and all of its first-tier suppliers. Their Lean Management is best known as Toyota Production System (TPS), which runs on two core philosophies: continuous improvement and respect for people. Some most notable components of TPS elements include (Toyota, n.d.):

1. Standardization to ensure a consistent approach to quality and a safe method of operation;
2. Complete elimination of waste from every process and activity;
3. Absolute elimination of overburden and unevenness in all areas to allow members to work smoothly and efficiently;
4. "Just-in-time" approach that was pioneered by Toyota and gained almost universal acceptance in world manufacturing.

"Just-in-time" is a term coined by a Toyota engineer. Implementing a just-in-time system means that Toyota "pulls" in needed products from its suppliers when a part needs to be replenished. Vendors of Toyota enjoy the same benefits of "just-in-time" of pulling needed products from their suppliers readily and efficiently exactly the same way it happens at Toyota's assembly plants. Product quality is improved as TPS includes metrics for signaling defects whenever and wherever they occur. The Part two—Library Supply Chain Model of the book emphasized the impact of collection service defects. One of the examples the author used was acquisition of substandard descriptive metadata or catalog records from vendors that accompany e-books or patron-driven acquisitions (PDAs) and the effect it has on user discovery and library circulation.

Perhaps, Lean is the most valuable philosophy for libraries today if it were to be implemented fully. Although PDA is a form of just-in-time practice, in general libraries routinely and largely operate on the opposite side of just-in-time concept. The current non-Lean management practices stifle people's imagination, creativity, and their sense of ownership. Most of the employees cannot gain insights of the entire Collection Supply Chain unless they step outside their own daily tasks and examine how their work relates to the tasks of others. Under Lean, each downstream function of the Supply Chain operates under the principle of "pulling" from the upstream function. Lean can create an environment where people have to think about what they do and what others do; why or why not their tasks or others' tasks relate, connect, and make sense; and thus bring continuous improvements to their day-to-day work along the Chain. Lean is a systematic approach to achieving the highest quality possible.

SPORADIC ENHANCEMENTS VERSUS SYSTEMATIC IMPROVEMENTS

Lean management is not a quick solution for cost reduction; it is a fundamentally different system from traditional methods of organizing and managing employees, suppliers, customer relationships, product development, production, and the overall enterprise (Blanchard, 2007). Liker and Franz (2012) suggested that Lean should be viewed as a system linking strategy to personal accountability. According to Liker and Franz, key factors that create achievable success are the following:

- A strong, committed leadership team that believes that teacher/coach model of people development is mandatory;
- An admission that problems, although painful in the short term, are learning opportunities and are valuable to the company;
- Seeing lean as an integrated system and not merely a collection of clever tools;
- Recognizing process improvements can only be sustained by raising the level of development of the teams that are using the process;
- A long-term commitment to learning and development to provide the basis for a continuous-improvement culture;
- People spending enough time in an area of specialty to develop deep functional knowledge and expertise; and
- Lean coaches who are respected teachers and not simply another "pair of hands" whose job is to solve your problems.

Cultural change and talent management are required for Lean to succeed. Because libraries are cultural and heritage institutions, long-term mind-set, consensus decision making, and continuous improvement should be ingrained into their DNAs. Based on that assumption, the author foresees a bright future for Lean taking roots in libraries. Where implementation of Lean strategies might run into issues, the libraries' not-for-profit characteristics are manifested in their loosely managed operations and historical despise of business economic values. The author suggests that libraries:

- Build transparent mechanisms to surface any collection problems in a steady-flow manner so that no problems can be hidden for any reasons;
- See collection holdings as inventory and thus apply inventory controls to avoid holding overages;
- Optimize and standardize collection processes to provide consistent experience (Chapter 15 suggests Perfect Collection Fulfillment as the result of proper collection services);
- Understand the value of support departments in the supply chain and match their capacity with collection demands;
- Respect vendors and partners and be cognizant of their business realities, but challenge and collaborate with them to improve the processes of the Extended Supply Chain;
- Take time to understand all options for subsystem improvements and impacts on the whole system, but implement the chosen options quickly;
- View Supply Chain processes as integral features of collection services.

Srinivasan (2004) recommended several principles on how to build and manage Lean Supply Chains. The author selected some that most apply to the library environment and annotated them below. Readers who are interested in Lean Supply Chain concept, techniques, and implementation details should read Srinivasan's book for more advice and suggestions.

1. Focus on customer needs and process consideration when designing a product (in this case, it will be collection services);
2. Buffer variation in demand with capacity, not inventory (at this point in time, libraries' capacity is tied to a large inventory of collections);
3. Use Forecast to plan and pull to execute supply chain (libraries should consider projecting collection demands and providing patrons with ideal search and delivery processes rather than accumulating and maintaining a large body of collections);
4. Build strategic partnership and alliances with members of the Supply Chain, with the goal of reducing the total cost of providing goods and services (under the context of the Extended Collection Services Supply Chain, libraries and vendors can eliminate much waste by unifying standards and protocols, streamlining processes, sharing core competencies, and embracing new technologies);
5. Design products and processes to promote strategic flexibility (Apply "Make-to-User" and "Make-to-Stack" models wherever fit to free up library budget and human resources);

6. Develop performance measures that allow the enterprise to better align functions and move from a functional to a process orientation model (increase collaborations across departments responsible for collection services and create metrics for improving process alignment);
7. Make decisions that promote a growth strategy and focus on improving throughput (create strategies to manage new collection demands and increase the speed of collection processing and fulfillments);
8. Synchronize flow by first scheduling the bottleneck resources on the most productive products, then schedule nonbottleneck resources to support the bottleneck resources (free up resources locked under a large inventory of collections to support strategically important emerging services, such as research support; the resources include more than just acquisitions budget, as the large body of collections require recurring costs of staff, equipment, facility, utility, etc.);
9. Focus on synchronizing the flow, not on balancing capacity (the flow of collection processes contributes much to the collection service excellence, so the capacity should be allocated to increasing the flow of collections to users).
10. Reduce variation in the overall system, which will allow the Supply Chain to generate higher throughput with lower inventory and operating expenses (implement standardization along the Supply Chain, as standards not only provide better options from economics standpoint but also deliver consistent service quality).

IMPEDIMENTS TO OR ENABLEMENTS OF SUPPLY CHAIN ADOPTION

Libraries must take a systematic approach to implementation of Collection Services Supply Chain and enablement of collaborations across internal departments and with vendors; otherwise, in no time they run a risk of sliding back into their old ways of doing business. Tension between change and stability is commonplace and expected. Libraries should keep in mind that Supply Chain is not a departmental solution. They should embrace the concept at the organizational level. Teamwork is critical to launching and sustaining continuous improvements. Senior management members need to guide its adoption down from their level to middle management and eventually all the way to individual contributors.

To become truly successful, libraries need to transition from doing cherry picking of Supply Chain models and tools to fully embracing SCM principles. In library industry, there is a disconnect between established strategic level of priorities and operational levels of practices. The truth is that real priorities are evident in institutional daily practices. Libraries will need to change their management style to get the complete benefits of Supply Chain Management on a sustainable basis. The change has to begin with managers who are willing to experiment and take some risks. Senior management needs to develop leaders who fully understand the benefits of the Supply Chain implementation, walk and walk according to the SCM principles, and train the rest of the institution's workforce to follow suit. The author suggests putting together a Supply Chain team composed of those leaders who will choose supply chain projects, document processes, and train each other on those processes. The team should focus on getting to the root causes of problems rather than on fixing symptoms of those problems.

LOOKING AHEAD

To summarize, the book advocates for a new type of relationship between libraries and their vendors. In many cases, the relationship currently in place may be considered strictly transactional in nature, so changes often feel as being pushed onto each other. For the betterment of user productivity, libraries and their vendors should consider improving the process, procedures, and practices collaboratively along with their Extended Collection Services Supply Chain. Systematic, scalable, sustainable, and mutually beneficial changes can only come from a real meaningful partnership.

The partnership should focus on long-term benefits while addressing short-term pains, but should not give in to immediate pressures. Too often, budgetary pressures, storage shortages, or even prestige or status in holdings may derail progress toward long-term benefits. All parties should only look for actions that allow sharing costs and benefits over the Extended Chain by contributing unique expertise to the overall goals. Those actions can be clearly defined as short-term milestones, which will lead to long-term benefits one step at time.

Any anticipated changes will require all parties to examine their current management philosophy, policies, procedures, standards, practices, and technologies. Currently, there are many areas that could be improved upon in relation to interoperability. The alignment of the above operational components will determine the productivity and efficiency levels of new partnerships and indeed yield benefits for all parties involved and their end users.

Finally, such partnership will have to manage the pushbacks originating from well-established traditional organizational structures, where complacency and bureaucracy may attempt to stall initiatives or progress of Supply Chain integration. The author wishes the best to those libraries and their vendors who are willing to labor together toward a goal of solving the growing challenges in the knowledge industry by viewing them through the lens of Supply Chain Management. At the end of the day, the Supply Chain is about improving user experience in new and more meaningful ways and setting a new standard of service excellence in the knowledge industry.

REFERENCES

Blanchard, D. (2007). Lean, green, and low cost. *Industry Week*, 37–39.

Burton, T. T., & Boeder, S. M. (2003). *The lean extended enterprise: Moving beyond the four walls to value stream excellence.* Boca Raton, FL: J. Ross Publishing.

Fawcett, S. E., Magnan, G. M., & McCarter, M. W. (2008). Benefits, barriers, and bridges to effective supply chain management. *Supply Chain Management: An International Journal, 13*(1), 35–48.

Keller, R. (2008). Silver bullets only exist in Hollywood: It's the slow, steady and continuous pursuit of excellence that wins the competitive race. (Continuous improvement). *Industry Week, 257*(10), 12.

Liker, J., & Franz, J. (2012). The Toyota way: Helping others help themselves. *Manufacturing Engineering, 149*(5), 87–95.

Rad, M. R. (2013). *10 characteristics of good problem solvers.* The Huffington Post. Retrieved from: http://www.huffingtonpost.com/roya-r-rad-ma-psyd/problem-solving_b_4302935.html.

Srinivasan, M. (2004). *Streamlined: 14 principles for building and managing the lean supply chain.* Mason, OH: Thomson.

Toyota Motor Corporation Australia Limited. (n.d.). *Toyota production system.* Toyota. Retrieved from: http://www.toyota.com.au/toyota/company/operations/toyota-production-system.

INDEX

'*Note*: Page numbers followed by "f" indicate figures, "t" indicate tables.'

T

U

V

W

Made in the USA
Monee, IL
15 September 2026

63317835R00109